CONTROL TECHNOLOGY

teachers' handbook

Control Technology was written by G.J. Fox and D.F. Marshall, Danum Grammar School for Boys, Doncaster, and was edited for Project Technology by G.L. Viles. It was revised by P.W. Ghee, D. Hendley, A. Paul and G.L. Viles (editor).

Project
Technology

CONTROL
TECHNOLOGY

teachers' handbook

HODDER AND STOUGHTON

LONDON SYDNEY AUCKLAND TORONTO

British Library of Cataloguing in Publication Data

Fox, G.J.
   Control technology.——2nd ed.
   Teacher's handbook
   1. Control theory
   I. Title II. Marshall, D.F. III. Viles, G.L.
   IV. Schools Council, *Project Technology* 629.8   QA402.3

   ISBN 0 340 36405 X

First printed 1975
Ninth Impression 1984
Second edition 1986
Second Impression 1987

Printed in Great Britain for Hodder and Stoughton Educational, a division of Hodder an
Stoughton Ltd., Mill Road, Dunton Green, Sevenoaks, Kent, by Page Bros (Norwich) Ltd

Contents — teachers' handbook

Foreword

Project Technology was a major curriculum development project initiated by the Schools Council to promote a better understanding by boys and girls in school of the importance and relevance of technology. The Project was concerned with helping teachers to stimulate an awareness of the material and scientific forces which effect change in our society and to develop knowledge of these forces and their means of control by the dirct involvement of pupils in technological activities.

The Project Technology teaching-material programme is the result of a careful assessment of what is required in the schools, followed by trials and editing of the material itself. The Project Technology team felt that it was essential to draw on the experience, imagination and flair of individual teachers who, over a period of years, had developed technological work in particular parts of the school curriculum.

It is against this background that the *Control Technology* course should be seen. Other teaching material, notably the Project Technology Handbooks series, indicates both a thematic and a tactical approach to school technology. All of this reflects the diverse nature of the work being done and of the alternative teaching methods and organisation which are possible.

This course, however, is intended to meet the needs of those schools who wish to develop a structured and sequential two- or three-year course, covering an important area of technology. Some teachers have expressed the view that, while appreciating the value of what might be regarded as the more fortuitous involvement of pupils with technological projects and investigations, they should welcome a more systemic course approach.

The joint authors of *Control Technology*, Messrs. G.J. Fox and D.F. Marshall, have developed the course over a number of years at Danum Grammar School for Boys, Doncaster, where they were given essential support and encouragement by the Headmaster (Mr E. Semper, OBE) and by the Doncaster Education Authority.

Clearly defined educational objectives were established, and the appropriate teaching methods, based on pupil assignments, with appropriate texts and equipment, were progressively developed, with the support of the Project Technology team.

First-stage school trials were conducted in selected schools in the Doncaster area with the generous help of the LEA. Secondary trials followed in various parts of the country. We are especially grateful to the teachers in these schools for the help and cooperation they provided for the authors, the Project Technology team (led by Mr G.L. Viles) and the Evaluators appointed by the Schools Council.

The *Control Technology* course has been carefully based on purpose-built equipment essential to the effective running of the course. A considerable effort has been put into the development of this equipment, at all stages, and we are indebted to all concerned, including the present suppliers. More recently, Trent Polytechnic are to be thanked for encouraging further development work by the National Centre for School Technology.

This revised edition has been prepared to serve the current examination syllabus set by the Associated Examining Board and the Joint Matriculation Board who at the time of writing, are the examiners for this subject at 16 + .

Assignments using an electric motor have been amended for use with an improved model. Electronic assignments now follow current practise in circuit design, and the pneumatic component symbols now follow standards in current use.

Fluidic control is not included in current syllabuses and this original section has been removed. The logic section has been extended to meet current syllabus requirements and symbol notation.

General Introduction

The Need for New Courses

The *Control Technology* course is part of the teaching material produced by Project Technology to enable teachers to stimulate an awareness of the material and scientific forces which effect change in our society. The course also enables pupils to experience the purposeful application of knowledge in an increasingly significant area of technology, affecting the working and domestic life of more and more people.

Work of a technological character is attempted in some schools through existing science and craft courses, but the work undertaken is often limited by syllabus requirements, especially in examined courses. An approach that will stimulate a greater awareness of the modern technological environment is needed. It must excite the imagination of children, including the applied scientists and engineers of the future. Although young children are naturally imaginative and inventive, the teacher's experience often confirms that this innate curiosity diminishes rapidly beyond the age of about 13 years. Many attempts to bring technological project work into the curriculum result in only modest success, since progress is often slow and can lead to frustration unless the student meets with frequent success. Too often success is achieved only after considerable time has been spent on constructional work or when the teacher has 'rescued' the project by giving excessive guidance and encouragement.

New courses for 13 year olds are, therefore, needed which will:

1   maintain and stimulate the inventiveness of pupils;

2   enable construction time to be kept to a minimum, thus giving confidence and encouraging experimentation with alternative solutions;

3   provide stimulating problems which will motivate children and require the minimum teacher contribution;

4   provide a rich variety of background information and techniques, enabling a number of solutions to be formulated to a given problem;

5   provide training and a logical approach to project work which will be especially valuable to the potential sixth-former;

6   be complementary to existing craft and science courses by emphasising the applications of scientific principles;

7   be flexible enough to keep pace with developments in technology;

8   allow children to work at their own pace, and also extend the brighter ones;

9   enable teachers without specialist qualifications to become involved.

The Control Technology Course

The course in *Control Technology* attempts to satisfy the above requirements. It was designed, in the first instance, for average and above average pupils who might not, be taking 16 + courses in physics, metalwork, or woodwork. Some schools have used the course with mixed groups of boys and girls.

Outline of the Course

The course is intended to last for three years, on the basis of two forty-minute periods per week in the third form and four periods per week in the fourth and fifth forms. A series of carefully programmed investigations and experiments, using purpose-designed equipment, should take approximately half the total time allocation. They are designed to cover the relevant subject matter and to provide experience in the solution of problems. The remainder of the time is to be used for projects of varying lengths. It will be appreciated that in the earlier stages of the course all of the time is given up to the programmed sequence of units and suggested mini-projects. Where necessary, important concepts are developed further by teacher demonstrations and discussions. The specially designed, easy-to-use, reliable equipment ensures rapid progress. In the later stages of the course the bulk of the time is given up to longer and more open-ended projects.

Some schools have successfully conducted the course on a two-year basis, but this inevitably means that less time is available for the major projects during the final year.

Educational Objectives

1 To provide sufficient background knowledge to enable pupils to solve complex problems involving the making, control, and automation of devices.

2 To encourage them to think creatively and to produce original, imaginative work.

3 To promote the ability to analyse what is involved in a new situation.

4 To promote the ability to select and apply principles, methods, and procedures, and to conceive probable solutions.

5 To promote the ability to synthesise and, thereafter, to plan and possibly to construct a device.

6 To promote the ability to recognise the limitations of a design, and to modify or suggest modifications to it.

7 To give pupils confidence in using unfamiliar and possibly complex equipment.

8 To encourage pupils to record their work in a clear fashion, including successes and failures.

Programmed Notes

Each programmed unit consists of:

1 pupils' assignments;

2 pupils' follow-up material, which,

    i)   provides possible solutions and relevant information,

    ii) confirms the fundamental principles involved,

    iii) prepares the ground for the next assignment,

    iv) gives background information.

The units are designed:

1   to allow pupils to work at their own pace, to extend their background knowledge, and to ensure that adequate information is given at the right time;

2   to enable teachers to give help to individuals or groups in difficulty;

3   to assist teachers who have limited specialist knowledge.

In addition, the extensive teachers' notes and Equipment Guide provide:

1   details of special demonstrations;

2   constructional details of special equipment;

3   sources of supply of mateiral;

4   suggestions for teaching method and modifications to meet various conditions, e.g. poorer ability pupils;

5   additional background material relevant to particular topics.

Syllabus

The course has been accepted by a number of Regional Examining Boards as a CSE Mode 3 examination. It has also been accepted by the Associated Examining Board and Joint Matriculation Board as a Mode 1 GCE O-level examination. During the period of school trials, and since, the course has been successfully conducted on both a two-year and on a three-year basis.

The following syllabus indicates the areas which are likely to be covered by the course, but should not be taken to indicate the sequence in which the ground is covered in the programmed units. Individual examination syllabuses may vary in detail and these should be consulted before commencing the course.

Structures. Tension, compression, bending, torsion, ridigity in two and three dimensions, moments, stability. Correct use of members. Use of strain-gauging and photoelastic techniques.

Rotary motion. Spur, bevel, helical, and worm gears. Chains and sprockets. Belts and pulleys. Gear trains (simple and compound). Effect of gear ratio on torque produced. Velocity ratio. Mechanical advantage. Friction. Efficiency. Electric motor. Characteristics of

4

the permanent-magnet d.c. motor; speed/voltage, load/current relationships. Work and energy as applied to the electric motor and mechanical systems.

Linear motion: Mechanical — Crank and slider. Cams. Eccentrics. Rack and pinion. Screw mechanisms.

Electrical — Solenoids and electromagnets. Effect of current and the number of turns. Effect of varying the position of the armature on the force produced by a solenoid.

Pneumatic — Single- and double-acting cylinders. Effect of air pressure and piston diameter on the force produced.

Basic electricity. Working knowledge of voltage, current, and resistance. Ohm's Law and simple calculations appropriate to control devices.

Difference between a.c. and d.c.

The transformer. Half-wave and full-wave rectification. Simple smoothing circuits.

Effect of current drain on the ripple of the rectified supply.

Protection devices — fuses, overload trips, earthing.

Use of electrical measuring instruments (voltmeter, ammeter, and ohmmeter).

Types, properties, and the use of capacitors — especially as applied to delay systems.

Series and parallel arrangements of cells, bulbs, resistors, and capacitors.

Introduction to electrical symbols and the diagrammatic representation of electrical circuits.

Resistance colour code, power rating, and tolerance.

Types of variable resistor — rheostat and potentiometer connections. Effect of temperature on the resistance of metallic conductors with special reference to the filament lamp. Effect of temperature on carbon and semiconductor materials.

Checking of components for serviceability — resistors, capacitors, coils, diodes, and transistors.

Switching — electrical. Introduction to the common types of switch and their uses.

Toggle switches, push switches, microswitches, reed-switches, wafer switches.

The differing poles and 'ways' of the various types of switch.

The relay — methods of determining the energising and minimum hold-on currents.

Heavy-duty and multi-contact relays.

The uniselector.

The use of relays to produce oscillators, delays, and 'flip-flops' The photocell as a switch.

Power dissipation in photocells.

Transistor circuits with special reference to the transistor as a switch.

The use of silicon diodes in switching circuits.

Switching — pneumatic. Manual, mechanical, electical, low-pressure and high-pressure operated valves. The use of valves to produce delays and 'flip-flops'.

Logic circuitry and applications. AND, NAND, OR, and NOR arrangements. Truth tables, logic equations. Design of logical control systems, schematic diagrams of logic circuits. Safety circuitry for the protection of a machine and the operator. The use of manual switches, relays, pneumatic valves and electronic logic units in the logical control of apparatus and devices.

Estimated Costs

The equipment and quantities specified are those recommended for groups of 15 students. Schools have found, however, that when a second and a third group follow on in subsequent years, some duplication of particular items of equipment is desirable. This further expenditure will vary with timetabling arrangements and with the type of project work undertaken in the final year.

(Detailed information on the equipment requirements for this course can be obtained from The National Centre for School Technology. Trent Polytechnic, Burton Street, Nottingham NG1 4BU.)

16 + Examinations

For a typical three-year examination course the continuous assessement can be provided by:

1 Set tests, under examination conditions:

    i) written — probably taken at the end of the second year and at the end of the first two terms in the final year;

    ii) practical — at the end of the course;

2 Course work, with special reference to:

    i) notebooks recording the results of investigational assignment work and mini-projects,

    ii) accounts of major projects.

Refer to examination boards for details of this assessment as it varies.

Distinctive Features of the Control Technology Course

The *Control Technology* course has been developed with the following considerations in mind.

Many of the more important aspects of technology can be taught through an understanding of control principles.

When a device is to be controlled, mechanised, or automated, a variety of restraints influence the final design. Typical constraints are the number of alternative solutions which can be considered in the light of previous experience, the ingenuity of the designer, the size of the installation, the cost, the speed of operation or repetition rate, environmental conditions such as excessive light or dust, types of sensor available, and so on.

The pupil working in the classroom or laboratory-workshop has limited previous experience. Free of inhibitions, he or she has to rely, more on his ingenuity and inventiveness. There are potentially a large number of alternative solutions to the control problems which can arise in the school course; consequently the teacher should encourage the pupils to accept the challenge of critically systematically examining each of their ideas. Because project work is a regular feature of the course, groups of more than 15 pupils will be difficult to handle by one teacher.

Young people are naturally inventive.

Many teachers feel that this inventiveness begins to decline rapidly at about 13 years of age. The reasons for the decline are probably complex, but for many pupils the decline may well be linked with the tendency to 'spoon-feed' in some examination subjects. This is brought about, some would say, by the syllabus requirements.

The demands made upon the pupils in the *Control Technology* course, together with the encouragement to experiment and devise solutions to problems, help to maintain the flow of ideas. Not only is the flow maintained, but the course encourages pupils to be practical and realistic in putting forward solutions. Indeed, as the pupils become familiar with the range of hardware available to them, and acquire an understanding of the relevant principles, their solutions more closely fit the specification.

The course should be flexible.

The structure and content of the course should reflect the fact that technology is essentially a dynamic process. The latest techniques and developments which are relevant to work at school level can be incorporated by the enterprising teacher, especially by way of minor and major projects.

The course has been written primarily for use by pupils of average and above-average ability. An interesting and stimulating course can also be provided for the less gifted by modifying some of the assignments, leaving out others and using suitable replacements, devising simple projects, and allowing more time throughout. It is suggested, however, that the course be used at first in its original form with able pupils, and subsequently modifications should be made. Teachers may tend to underestimate, at first, the standard of work attainable by their pupils: trials in various schools have shown that, with sufficient motivation, high standards are possible with pupils of lower ability.

The normal sequence of work for a three-year course* would be:

1st year
- Structures   (including two mini-projects)
- Gears
- Basic electricity
- Electrical switching
- A minor project — based on work to date

2nd year
- Linear motion
- Pneumatic control
- Electronics
  - — resistance
  - — rectification
  - — the transistor
- Logic circuits
- A project — of a half term or one term duration

3rd year
- Any outstanding assignments
- A major project — one to two terms duration

Detailed guidance and suggestions are given in the teachers' notes for individual sections of the course. For example, it may be desirable for some pupils to omit certain assignments, or to postpone them until others have been completed; alternatively, some teachers might wish to give greater emphasis to the work on Structures, for instance, by encouraging children to construct models from balsa wood and to test them to destruction.

Being sectionalised, parts of the course may be taken to enrich the work in other subjects. Teachers have already found that, in addition to teaching the *Control Technology* course proper, they can also use some of the methods and equipment to advantage in other areas, particularly on a modular-course basis.

The course must fit its objectives.

Whether the course is used as it is presented or in some modified form, it is hoped that teachers will have certain objectives in mind. The course has stated objectives and many teachers will accept these, since they form a firm basis for the work in all its aspects.

---

*Reference is made later in these notes to the possibility of a two-year course structure.

Factors to be considered when teaching a modified version would include:

1 the relevance of the topics of the course to the stated objectives;

2 how well do the teaching methods used for a particular topic fulfil the objectives?

3 the relative importance of each objective as judged by the teacher;

4 is an objective determined by consideration of general education or vocationally-directed education?

5 How much emphasis is given to each objective

    i) by virtue of the course content?

    ii) by virtue of the relative frequency of the methods used to teach the various topics?

The equipment should be reliable, easy and rapid to assemble, attractive, and compatible.

The electrical equipment is housed in attractive units; electrical connections being made through 4 mm sockets and stackable plugs. The soldering of wires, except in major project work, has been avoided in order that circuitry can be made up quickly. By using the units, with connecting leads, quite complex circuit arrangements can be set up in minutes. Should a design fault or limitation become evident, the arrangement can be dismantled and an alternative method tried again in a very short time.

All the electrical equipment has been selected in order to be operated from 12 volt d.c. supplies. When alternative equipment to that recommended is being considered, teachers should ensure that the units will operate under all the required conditions, as indicated in the assignment sheets. For instance, a different photocell may not operate the relay unit or, alternatively, a different type of relay may not operate from the photocell satisfactorily under the conditions assumed in the assignment. Any new units introduced should also operate satisfactorily on 12 volt d.c. supplies.

Surplus equipment — industrial and military — is a useful source of 'hardware' and components for projects in the Course. There may be a temptation to use complete units but, whilst occasionally this may be desirable, in general such equipment is more valuable as a source of 'bits and pieces'. Assuming that a suitable store is available, this equipment is probably best left intact until specific components are required by pupils for a major project.

Teaching Methods

The subject matter of the course is generally taught by a combination of four different methods:

1 programmed or sequential assignments and follow-up sheets;
2 teacher demonstrations;
3 homework questions and assignments;
4 project work.

Programmed assignments.

The assignment method of teaching is used for the following reasons.

1 It ensures that all children cover the same basic material.

2 It enables pupils to work at their own pace. Some pupils obviously work faster than others and, since most of the principles are learned through experiment and investigation, adequate time must be available to allow for different rates of working.

3 The follow-up sheets contain additional information, some of which is 'for information only' and would not warrant discussion time. In effect, some of the information in the follow-up sheets is used as one uses a textbook. The majority of teachers involved in school trials, however, felt strongly that the follow-up sheets should not be made available to pupils until they had spent sufficient time working unaided on the assignments.

The teacher should visit each group of pupils as they work, to discuss and to check progress. He should ensure that full answers are written, as indicated by the questions and statements printed in bold italic type in the assignments, and also that suitable notes are made from the follow-up sheets. These notes should include points which have not been understood as pupils have worked through the assignments and also the important additional information not referred to in the assignment.

Should pupils make errors as a result of misunderstanding of basic principles met previously, they should be encouraged to refer back to their notes or to the appropriate assignment. All pupils will meet difficulties from time to time, and the advantage of making suitable notes will quickly become apparent to most. The teacher should avoid providing complete answers, but should direct pupils to the source of the answer.

Teacher demonstrations.

A number of these are covered in the teachers' notes, but as the teacher becomes familiar with the course he will no doubt devise others. Although buffer experiments are suggested from time to time, in order to extend the faster and brighter pupils, it is possible that timing difficulties will still be experienced with mixed-ability groups. Should too wide a gap arise between groups or individuals, the teacher should use this discretion and demonstrate certain assignments to help bring the group level. Such demonstrations may also be necessary because of a shortage of apparatus on occasions.

Homework assignments and questions

Many schools will allocate some homework time to pupils who are taking the *Control Technology* course. Such time is valuable, for it allows pupils to work through assignments in class and, for homework, write up the notes and/or answer homework questions.

Project work

In the early stages of the course, pupils require considerable guidance in project work. The initial projects, therefore, occur as part of the assignments, and pupils tackle the problems in a planned sequence. In later project work it is recommended that pupils are reminded, from time to time, of the notes *The approach to project work* at the end of the assignments book. Four different types of projects are suggested for inclusion in the assignments.

1 *Mini-projects.* These are the projects specified in the assignments, e.g. the design and construction of a bridge, or the design and construction of a lifting device. In this kind of project, a suitable specification is given and the solution is

closely related to recently completed assignments.
Materials — mainly Meccano and the basic electrical units.
Time required — two or three double periods. (Two or three
pupils per group.)

2 *Minor projects.* These projects are usually attempted after the
Electrical-switching assignments. Pupils are given a list of
possible project titles (not a strict specification) from which
they choose one. Alternatively, if they wish, pupils can suggest
their own subject. Ideally, the problem and the solution should
be related to all the assignments completed up to that time.
When all the projects are completed, it is helpful if a member of
each group describes and explains their project solution to the
others. Materials — primarily the quick-assembly units. Time
required — four to six double periods. (Two or three pupils per
group.)

3 *Major projects.* These are usually attempted after the *Linear-
motion* and the *Pheumatic-control assignments.* Here the same
specification might well be given to all the groups. Each pupil is
asked to consider several solutions and to select the most satisfactorily; it is then possible to group together those pupils who
are hoping to make use of similar principles or methods. The
pupils in the group then select the best solution or combination
of solutions. Materials — a wider range of materials, including
Perspex, wood, Meccano, and any control devices. Pupils
should be discouraged, perhaps, from using only the quick-
assembly units. Time — between half a term and a full term.
Where the major project is used for examination assessment,
some examination boards prefer this to be the work of
individual pupils.

The Teacher's Role in Project Work

It cannot be emphasised too strongly that in this course all projects
should be primarily the work of the pupils and *not* that of the
teacher. When organising the project work, the teacher has,
perhaps, three main functions:

1 to act as a supplier of material, equipment, and components
appropriate to the project;

2 to provide a stabilising influence throughout, and to ensure that a project is feasible, technically and financially;

3 to provide advice and guidance on which pupils can rely, particularly when they are unsure of basic principles.

The teacher's main function, then, during the design and construction of a project, is to check that principles are being applied correctly. If he is doing much more than this, then the pupils could become, at worst, mere labourers for the realisation of his ideas. As the project is completed, it should have developed from the pupil's ideas. This must not be confused, however, with the fact that sometimes a teacher may have made part of a device. If he has done so, it should have been at the request of the pupil and to meet a specific need.

Where does Control Technology fit into the Curriculum?

The course is designed primarily for boys and girls of average and above-average ability. It will normally be of three-years duration, with pupils taking the course from 13 to 16 years of age. Based on a timetable of 40 minute periods, a typical time allocation would be:

Year 1   one double period;
Year 2   two double periods and one single period
          (or one triple);
Year 3   two double periods and one single period
          (or one triple)

Some schools have found that a two-year course is possible, since most of the third year is devoted to a major project. In this case, more emphasis should be given to projects in the second year, possibly at the expense of some of the assignments.

The *Control Technology* course was not intended to be an alternative to either science or craft work, but has been developed as a course which is complementary to both. Although it is not essential, it is desirable that pupils have some basic knowledge of physics and of craftwork before starting the course.

For convenience in some schools, *Control Technology* has been fitted into craft or technical-studies option schemes, or even into craft-plus-science option schemes, causing little disturbance to the rest of the curriculum.

Teachers with an interest in technology are usually to be found in physics or technical departments, and it is, therefore, mainly from these disciplines that the teachers of Control Technology emerge. Experience has shown that ideally there should be at least one member from each department willing to contribute, since both will have specialised knowledge and experience of value.

Classroom, Workshop, or Laboratory?

A room with the following facilities is required.

1 An adequate number of mains electrical power sockets,

2 a bench or pillar drilling machine,

3 a portable electric drill,

4 two sets of twist drills (limited range),

5 at least five 12 volt d.c., 2A power-supply units.

The following are also recommended.

6 Storage space for 'ex-equipment' items and components etc.,

7 storage space for unfinished projects,

8 storage space for quick-assembly units, Meccano, pneumatic equipment, etc.,

9 an air compressor,

10 storage space for paper, books, and miscellaneous items.

Project Suggestions

Minor projects undertaken after Electrical-switching assignments

1 Various burglar alarms.

2 Bridge which automatically lifts as a vehicle approaches.

3 Warning device for an over-weight load approaching a bridge.

4   Device for removing articles of above standard height from a conveyor.

5   Automatic flashing light for given variable lengths of time.

6   Device for counting the number of vehicles crossing a bridge.

7   An automatic 'scarecrow'.

8   An electrical lock for a safe or cupboard.

9   Safety devices to be fitted to a fairground roundabout.

10 Rain detector.

11 Sorting machine for coloured balls.

12 Traffic-light control system.

13 Automatic turntable for model locomotive.

14 Automatic level-crossing gates.

15 Liquid-level indicator.

16 Device for counting the number of people entering a room, but not those leaving.

17 Circuit for a quiz game, to indicate which team first pressed the button.

18 Automatic garage door.

Some Suggestions for Major Projects.

Warning devices:	'Low bridge' warning system. Motorway speed indicator.
Dispensing devices:	Delivery of correct quantities — e.g. marbles, nuts and bolts, liquids.
Sorting devices:	Selecting objects of different width, length, or colour. Selecting components according to their properties, e.g. selecting good or rejecting faulty diodes.
Repetition Devices: (sequence switching)	Drilling holes in blocks of wood. Drilling equally spaced holes in metal strip.

Repetition devices: (programmed)	Drink dispenser.
	Automatic gantry crane.
	Automatic metal-plating equipment.
Display devices:	Lighting of bulbs to indicate a sequence.
	Lights indicating information.
	Electrical indicator board for school timetable display.
	'Random' light display.
	Digital display for scoreboard.
'Reaction' devices:	Automatic control of a swing bridge.
	White-line follower.
	Quiz-game reaction indicator.
	Constant-level reservoir.
Selection devices:	School 'pupil-finder' punched-card-operated display.
	Telephone exchange for school.
	Equipment for the disabled.

*All of these projects have been undertaken by pupils involved in *Control Technology* examination courses.

The Control Technology Approach to Project Work

1   All pupils work through a programmed basic course which develops the ability to solve problems with little help from the teacher.

2   Progressive training is given in the approach to project work (see note on the types of project).

3   Quick, easy to use, and reliable equipment allows pupils to test their ideas before time-consuming practical work is undertaken.

Using this equipment, pupils gain confidence and achieve regular success.

4   Quick-construction methods are used whenever possible, e.g. basic electrical units, Meccano, etc. This allows a number of modifications to be made, if necessary, in the minimum of time.

5 The content of the basic course permits many stimulating projects to be set — projects naturally grow out of the course.

6 The solution to the project is initially unknown, with the pupils having to start from first principles. Solutions are rarely to be found in textbooks.

Some Problems Encountered in Project Work

1 The difficulty of choosing a suitable topic.

2 The pupils' limited knowledge. Difficulty may be encountered in locating sources of information, resulting in unfortunate delay. The information may be too advanced for the pupil to understand or apply.

3 Availability of specialist techniques, equipment, and materials.

4 Pupils' lack of experience in the whole approach to project work.

5 Some project work involves little problem-solving — the pupil may begin with a pre-conceived solution.

6 The initial stages of some projects are the most difficult, because decisions taken in the first few weeks often determine later work.

7 A team-teaching approach is often desirable when knowledge and guidance in various fields are necessary.

8 Repeated failure or frustration can affect morale. A measure of success is normally required.

The Teacher's Role in Project Work

1 The project must be primarily the pupil's work and not the teacher's. Too much intervention by the teacher can seriously reduce the educational value of this portion of the course.

2 When pupils need correction, the teacher should correct, if possible with regard to principles only, by referring pupils to their previous assignment notes.

3 The teacher should ensure, as far as possible, that:

i) the project is a suitable one;

ii) the proposed solutions are feasible, in order that pupils avoid having to make too many unsuccessful attempts.

4   The teacher should provide advice and guidance, together with appropriate resources — components, materials, equipment, etc. — when asked.

The Recording of Project Work

The thorough recording of project work is important:

1   because pupils will need to refer back to their previous work, particularly when problems are encountered;

2   for examination assessment purposes. The written account is often all that exists of the project, which may have been dismantled after completion and testing.

The following guidance may help in the recording of project work.

1   It is essential that a record is made as the work progresses.

2   A range of alternative solutions should be considered. This advice must be stressed, otherwise pupils will be content to develop their initial, often mediocre, ideas.

3   The reasons for adopting a particular solution should be noted.

4   When modifications are needed later, the reasons for these should be given. This is often the only way of knowing that other work was carried out, once the project is completed.

5   A concluding section should indicate how well the project fulfils the specification. When a project is only partly successful, an investigation into the causes of failure should be carried out and briefly recorded.

6   Precise recording is essential. The use of appropriate line-diagrams, freehand sketching, and photographs should be encouraged. Descriptions can be in concise note form.

The adequate recording of project work is largely a question of habit; therefore guidance should be given by the teacher to ensure that suitable recording is carried out as the work progresses. A little time at the end of most lessons can be used to record the progress made.

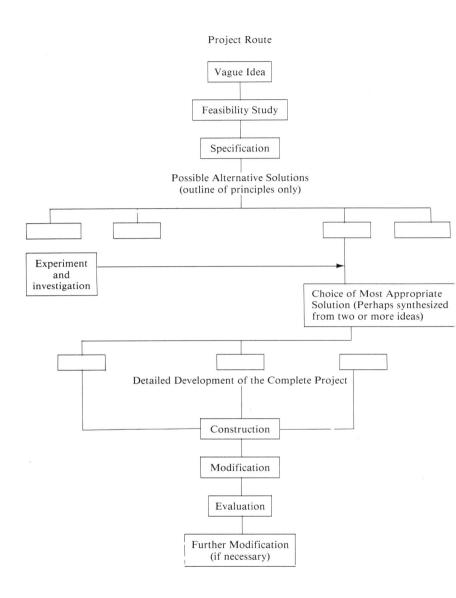

Project Route

Vague Idea

Feasibility Study

Specification

Possible Alternative Solutions
(outline of principles only)

Experiment
and
investigation

Choice of Most Appropriate
Solution (Perhaps synthesized
from two or more ideas)

Detailed Development of the Complete Project

Construction

Modification

Evaluation

Further Modification
(if necessary)

The Approach to a Project

In the early stages of many projects, the pupils may start with
rather vague general ideas and suggestions — for example, special
equipment for use by the disabled — then, after a period of

research, they should formulate their own appropriate project specification.

The following example, however, has been chosen to illustrate the manner in which a wide range of solutions should be considered in many *Control Technology* projects. The specification might be worded in this case, as, 'To design and construct a dispensing machine to place, repeatedly, ten marbles into suitable containers.

The device would appear to require four distinct parts or units:

1　a marble-feed mechanisms;

2　a method by which the correct number of marbles is selected;

3　a system to deliver the marbles to the container;

4　the provision of containers in the right place at the correct time.

If each problem is broken down into its essential parts, a number of different ideas can be developed for each part. (Ideally each solution should make use of a different principle, rather than involve the minor modification of a previous idea.)

Some possible solutions.

1　Marble Feed Mechanism

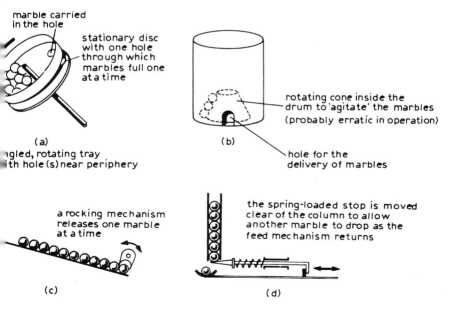

marble carried
in the hole

stationary disc
with one hole
through which
marbles full one
at a time

(a)

angled, rotating tray
with hole(s) near periphery

rotating cone inside the
drum to 'agitate' the marbles
(probably erratic in operation)

(b)

hole for the
delivery of marbles

a rocking mechanism
releases one marble
at a time

(c)

the spring-loaded stop is moved
clear of the column to allow
another marble to drop as the
feed mechanism returns

(d)

2 Quantity-selecting device

a) By weight

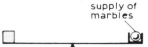

supply of marbles

with the correct quantity of marbles the balance will tip (only reliable if the marbles have very similar masses)

b) By measurement of a column

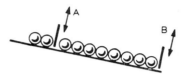

stop A moves down
stop B lifts to release the correct quantity
stop B returns
stop A lifts to allow more marbles to pass
(only reliable if the marbles have very similar diameters)

c) By counting

each marble is counted individually; when the correct quantity has passed, the feed is cut off while the container is replaced

3 Delivery system

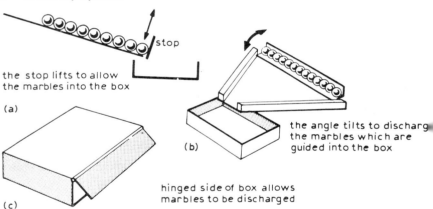

stop

the stop lifts to allow the marbles into the box

(a)

(b)

the angle tilts to discharge the marbles which are guided into the box

(c)

hinged side of box allows marbles to be discharged

4 Provision of containers

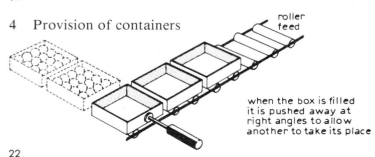

roller feed

when the box is filled it is pushed away at right angles to allow another to take its place

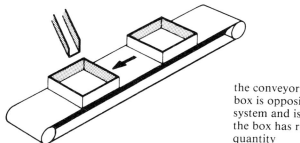

the conveyor belt stops when the box is opposite the delivery system and is restarted when the box has received the desired quantity

Each proposal should be analysed and evaluated (this often requires the building of models or 'rigs' to test ideas) and the most suitable solution to each part is selected and the reason is given.

The choice may be modified when the system is examined as a whole. One selected method may be best operated electrically, and another pneumatically; generally it is advisable to limit the power sources to be used on grounds of cost and convenience. Obviously it is essential that all the parts in the final design are compatible.

Each idea must now be developed. Ideally, detailed drawings should be made before construction work is commenced; however, some pupils have difficulty in expressing three-dimensional forms on paper, and interest may decline if the teacher always insists on suitable drawings. Indeed, in some circumstances it may be better to let pupils start constructional work without detailed drawings, when working with materials and equipment which allow modifications to be made as the work progresses. It is likely that modifications will have to be made during the construction and testing stages. These changes should be recorded and explained.

The pupil's evaluation of his project is most important; a critical appraisal should be attempted whenever possible. Avoid undue discouragement from the project which has failed: an analysis of the reasons for failure, and perhaps an attempt to correct the faults, can be of more value than the project which is only marginally better but, since it performs satisfactorily, is not viewed as critically.

In a group project it is important that all pupils are involved in major decisions and have an overall understanding of the project. To disseminate information on the details of a project, the use of an overhead projector has been found helpful. Each pupil is able to outline his individual contribution, and others in the group can then extract the information they consider necessary to produce an adequate project write-up.

Guidance notes

Introduction

The following notes are intended to be used in conjunction with the pupils' materials for the *Control Technology* course. An effort has been made to lighten the teacher's burden, with regard to preparation for the teaching of each programme, by dividing the notes into clearly defined sections. It is hoped that, before introducing a particular programme to his pupils for the first time, the teacher will carefully read through each section of the notes. These notes will give an overall picture of what is involved in the relevant programme and how the programme relates to those which follow. Instructions on the construction and use of demonstration apparatus are included.

The following five aspects are usually covered.

1   Guidance notes. These include the aims to be considered for a particular programme. Some of the material and a number of the explanations given in the Teachers' Notes are intended primarily to be of benefit to the teacher; however, a capable group of pupils might be able to cover some of the extra work, in which case periodic class discussions could be fruitful.

2   A list of apparatus required for fifteen pupils, giving five groups of three.

3   Demonstrations which might be used preceding or following the actual programme. The equipment and its use is described, but constructional details are given separately, where appropriate.

It is not anticipated that all demonstrations will be given quite as suggested. The teacher is expected to select and to use only those demonstrations which he feels will be effective. In some circumstances the teacher may decide to present the problem to the whole group, or alternatively he may set the challenge to a competent group who will later demonstrate their findings.

4   Buffer experiments. The use of programmed notes reveals the need for buffer experiments, since bright pupils will obviously work faster than others. In most cases, these experiments will be extensions of those suggested in the programme.

5   Homework questions. A period of homework is often

desirable, even if this consists of recording the work done in class. Some teachers may welcome typical questions appropriate to the programme.

NOTE: In order to assist pupils (and teachers) in determining when work should be recorded in notebooks, certain sentences in the assignments are printed in bold italic type. This is intended to indicate that a permanent record should be kept. This method is not used in follow-ups, since it is anticipated that pupils will record anything mentioned in them which they have not already fully noted or understood.

Structures

The main purpose of this section of the course is to enable the pupils to use materials intelligently and efficiently. We cannot assume that pupils will initiatively be able to use Meccano to produce rigid structures. Briefly, the programme is arranged to introduce to the children some ideas about triangular formation of members, tension forces, compression forces, the forces involved in a member which is under bending, and the stability of structures. Such information is needed, since, although we need not place undue emphasis upon craftmanship, the rigidity of a structure is vital, particularly when a device containing moving parts is to function satisfactorily.

Structures Assignment 1

Here we are asking the children to fasten together four pieces of Meccano strip and to test the arrangement for rigidity in one plane. (It should be remembered that when we ask, 'Is the structure rigid?', we are concerned with only two dimensions. Obviously we should not discourage anyone who wishes to extend this to three dimensions.)

Several solutions apart from fastening the strip across opposite corners will be suggested, but we should emphasise that triangulation provides the best solution.

At the end of *Structures follow-up 1*, the pupils are asked to consider which kind of brace is the most suitable for producing a rigid structure. After allowing sufficient time for thought, possibly by waiting until the next lesson, the teacher could:

1 demonstrate the best solution (see section on demonstrations);

2 explain why one solution is preferable to the others. Suggestions for further demonstrations are also mentioned.

Follow up 1 shows the Warren girder, and teachers should at this stage have available clear photographs or drawings to illustrate that the arrangement is indeed used extensively for bridges, cranes, and other structures.

A period of discussion should follow, in which questions of stability and moments may well be put forward by pupils as an introduction to work which follows later. In any discussion arising out of the programme, we should avoid giving too many answers but should leave some matters for further consideration.

Apparatus required (15 pupils, 5 groups of 3)

The quadrilaterals can be made from any lengths of Meccano *flat* strips. At least 20 pieces are therefore required initially. Additional lengths of flat strip, corner brackets, and gussets will also be needed for making the structures rigid.

Hence the apparatus required is:
    assorted Meccano flat strip,
    some corner brackets (part no. 133),
    Meccano nuts and bolts,
    means of applying forces (other than by hand), e.g. string, spring-balances, and weights.

One of our aims in this course should be to make pupils think and formulate their own solutions to problems. It is easy to present the solution by making a statement or suggestion at an early stage, or by providing a set of components carefully selected to solve the problem.

To avoid the latter, the following alternative suggestions may be useful regarding the supply of components during a lesson.

Give pupils free access to a range of Meccano parts. Let them select any component which they wish to use to solve their problems.

Even if it is necessary to limit the amount of Meccano for the lesson, ensure that, in addition to those parts required to solve the problem, other Meccano parts are included so that a certain amount of selection and rejection is required. Flat strips and nuts and bolts to make the quadrilaterals could be made available at the beginning of the lesson, whilst other parts — such as more flat stips, corner brackets, and gussets — could be withheld until the design work has been completed on paper.

Demonstrations

Photographs of bridges, cranes, etc., showing the use of triangulation can be introduced at this stage. These might be followed by the four demonstrations to show the best bracing method.

Demonstration A

Construct a framework as shown below.

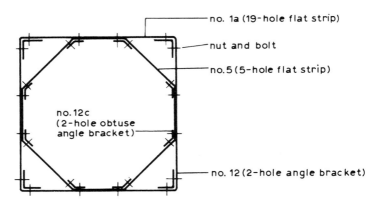

Apply forces, by hand, at the opposite corners (i.e. push or pull). The corners will retain their original shape, but the long side members will bend easily.

The amount of bending will depend upon a number of factors, some of which can be difficult to explain, and teachers may wish to study texts on Structures to learn more about the subject.

The following diagram shows the distorted structure and the forces in the individual members.

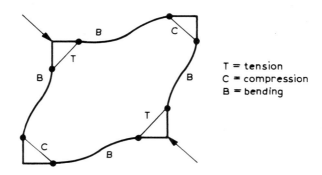

T = tension
C = compression
B = bending

It is not suggested that the kinds of forces existing in the various members should be discussed fully with pupils, since this is covered later.

Demonstration B

Construct a framework as shown below.

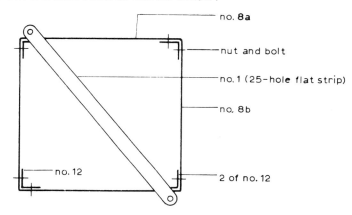

no. 8a

nut and bolt

no. 1 (25-hole flat strip)

no. 8b

no. 12

2 of no. 12

Apply forces to W and Y as shown.

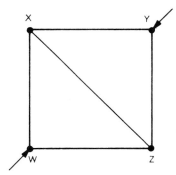

Very much larger forces are now required to produce any appreciable distortion of the side members, showing this to be a much better solution. The explanation for this is that all the members in the framework are either in tension or compression only. Bending of side members occurs as a result of excessive compressive forces. The forces in this structure are along the length of a member, and only when the compressive forces, in a 'perfectly' straight member, cause dislocation in the structure of the material will bending occur. (In practice a member is never quite 'straight' and bending occurs with modest forces.)

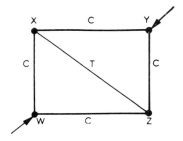

Demonstration C

The following demonstration will emphasise the difference between bending in pin-jointed structures and bending caused by forces other than those applied along the length of a member.

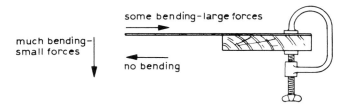

Hold a length of flat strip at one end by clamping to a bench, allowing most of the length to overhang. Show that (a) no amount of 'pulling' (tension forces) will cause bending (b) very large 'push' forces (compression) are needed to cause bending, and (c) a finger pushing at right angles to the strip easily causes bending.

Here we are simulating what happens in the following:

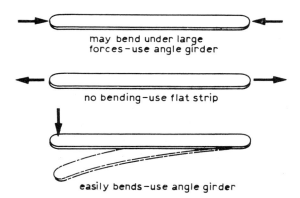

Demonstration D

From photoelastic material, cut strips 12 mm wide (take care not to stress it by clamping etc. — see *Appendix 1*) and construct two right angle supports as shown.

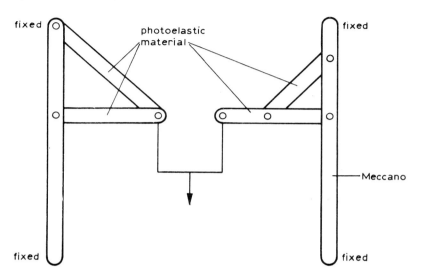

This demonstration is most effective if both arrangements, when similarly loaded, can be viewed simultaneously. This is possible with the apparatus described in *Appendix 1*.

Special equipment — photoelastic stress analysis apparatus (for construction details and brief notes see *Appendix 1*).

It may be helpful to use this apparatus to provide further evidence that the solution shown in Demonstration B is better than that shown in A. As the apparatus will be used again in *Structures 3* to demonstrate the force acting in a loaded beam, Demonstration D could be delayed until then and used to recall and reinforce the work of the first section.

Possibly the easiest way to explain the pattern formed in photoelastic material when stressed is to compare the lines (fringes) with the contour lines on a map. Where there are many lines close together (mountains), the structure is highly stressed; if there are few lines, the stress is low; if the colour is the same throughout, the stress is evenly distributed.

On loading the two structures, more fringes will appear in the structure on the right — particularly in the horizontal member, as this is in bending. The short supporting member is also highly stressed in tension.

In the pin-jointed structure on the left, the members are less highly stressed and are in either compression or tension.

Buffer Experiments

1   Repeat the programme using an entirely different shape of structure, and compare the results with the first shape.

2   Take measurements of the forces needed to distort the structure, with and without reinforcement, using spring-balances.

Homework Question

1   Examine the construction of electricity pylons, cranes, and bridges to see how the principle of triangulation is applied. Make sketches in your notebook.

Structures Assignment 2

We now approach the problem of three-dimensional structures. The pupils are asked to construct a box-like structure entirely from flat strip. They are asked to check the rigidity of the structure. They will notice that the vertical sides will bend either inwards or outwards under a vertical load, and that any force sideways will cause distortion of the shape as the members pivot about the securing bolts. With their knowledge of triangulation, pupils should then attempt to achieve some rigidity. At this stage they must be aware of the various Meccano parts available. Angle girders could be used for the vertical members, to give greater rigidity.

We should emphasise here, and elsewhere in the course, that when designing full-size structures other factors besides strength have to be considered. These include the weight, transport and assembly problems, and the cost of the final product.

The programme does not suggest methods of loading the structure when it is being tested, so we might encourage them to use their ingenuity. Whilst no given load is specified, when completed, the *improved* structure should bate least be capable of taking the full weight of a standing pupil. The 'balancing act' will inevitably produce sideways forces as well as vertical; obviously the teacher must keep a careful watch to prevent serious damage to parts.

Apparatus Required (five groups)

For the construction of the tower illustrated in *Structures 2*, the following Meccano parts are required:

an assortment of flat strip (at least 20 long strips and 20 short ones);
double angle strips;
Meccano nuts and bolts.

The use of long supporting strips (part no. 1 — 320 mm) indicates bending most effectively.

As suggested in the notes, these vertical strips should be replaced with angle girders if a really satisfactory solution is to be obtained.

Other parts required to carry out modifications:
long angle girders;
assorted flat strip or angle girders for cross-bracing.

Demonstrations

1 Photographs of tower cranes etc showing their construction.

2 The teacher should construct a tower showing an acceptable solution. He should demonstrate that it will withstand a pupil's weight and perhaps his own.

Possible solution:

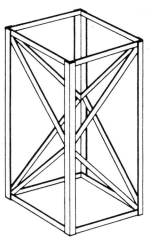

Buffer Experiments

Devise other methods of applying and measuring forces. Some pupils could undertake a quantitative analysis.

Homework Questions

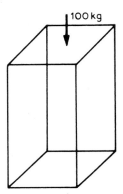

1   The diagram shows a tower (not very rigid) made from strips of steel bolted at the corners. A platform is placed on the top of the tower and a 100 kg load is applied centrally as shown.

 i)   Are the vertical members in tension, compression, or bending? Explain your answer.

 ii) If the force was increased gradually, what do you expect would happen to the vertical members?

 iii) Are there any forces in the eight horizontal members and, if so, are they forces of tension or compression?

 iv) If you had a means of measuring the forces in each of the vertical members, what force would you expect to find in each?

2   The diagrams (a), (b), (c), and (d) show a simple tower with the vertical members made from flat strip. A platform is placed on top of the tower, and forces are applied in the directions shown by the arrows.

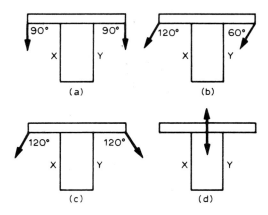

i) Assuming that the two forces are equal, say what you would expect to happen to the two vertical members X and Y in each of the diagrams.

ii) In diagram b, the two forces shown could be replaced by two forces at right angles to one another (not parallel to one another as shown) and still have a similar effect. Draw a diagram showing these two new forces.

Structures Assignment 3

This mini-project, which may take two to three lessons to complete, enables pupils to put into practice what they have discovered in *Structures 1* and *2*. The specification given may be changed to suit the available materials. The important part of the project, however, is to restrict construction to the use of lengths of Meccano which are short compared with the bridge span itself, thus ensuring that triangulation and the correct kind of section is used.

A fast-working group could investigate the forces acting on the ends of the bridge (reactions). This should be followed by a demonstration to show the principle of moments.

Before embarking on the project, it is helpful to suggest to pupils an approach to project work to ensure logical procedure. The following is offered as a guide and appears in the *pupils' assignments* book.

The Approach to Project Work

The bold italicised words indicate the need for a written record of the work.

1   *State* the problem precisely.

2   *Outline* several possible solutions.

3   Choose one solution and *say why* you chose it in preference to any other.

4   Divide the problem into several separate sections. *Work on each independently*, keeping in mind that all sections must fit together.

5   With the aid of *sketches* make a plan of how you intend solving each section of the project.

6   *Suggest* any modifications during or after construction.

7   Test the items which have been constructed. *Say* where your design has succeeded — and how well — and where it has failed.

8   *Suggest* further modifications for improvement. These may or may not be carried out.

The above procedure helps to illustrate that success can follow initial failure. We encourage pupils to have a positive attitude by analysing the situation to discover the reasons for failure.

It is recommended that the pupils should keep fairly rigidly to a pattern of this sort, otherwise poor quality reports will result. As the course progresses, it is hoped that pupils will also realise the importance of a written record. Later work will emphasise this point, since they will find they cannot progress rapidly unless they can refer back to previous work, particularly circuit diagrams. The procedure will encourage them to approach a problem logically, thus eliminating guesswork.

Pupils will need considerable guidance to enable them to express their thoughts visually. Sketches should be of a simple nature, so that ideas can be shown quickly and clearly. It is felt that engineering-drawing techniques are inappropriate because:

1   additional special equipment (and space) would be required;

2   progress would be slow because advanced techniques would have to be taught, and much time would be spend in making accurate drawings of ideas, many of which would be rejected before the drawings were completed;

3   very sophisticated techniques are needed to represent satisfactorily the complex arrangements developed in later work.

However, simple line diagrams are not always satisfactory, as they become very confusing when showing complex three-dimensional shapes.

Apparatus Required

In addition to a variety of Meccano strip, angle girder, and nuts and bolts, some strip material would help to make the bridge deck more realistic.

A simple design will need approximately twenty to thirty pieces of Meccano. It might be advisable to check the number of parts available before starting this programme, and if necessary reduce the number of groups.

Special Demonstrations

1   Demonstrations to show the forces acting in a loaded, end-supported beam.

i)   *Rubber-model demonstration*
A block of rubber approximately 225 mm long, 50 mm wide and 25 mm thick is used.

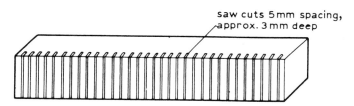

saw cuts 5mm spacing, approx. 3mm deep

When loaded, the rubber bends as shown. The distortion of the saw cuts clearly demonstrates the tensile and compressive forces present. The cuts in the lower half open up considerably, indicating tension; and in the upper half the saw cuts close up, showing compression.

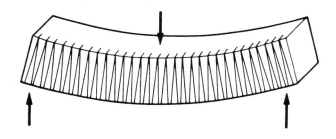

ii)   *Photoelastic stress analysis demonstration*
Equipment required: photoelastic analysis apparatus (*Appendix*);

strip of photoelastic material approx 225 × 12 × 3 mm; four 4 mm diameter holes drilled as shown.

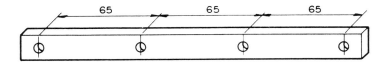

By applying the load from the two points A and B, circular bending is obtained.

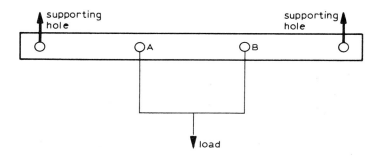

When the strip is loaded, a dark line appears approximately along the centre line of the strip (this indicates the neutral axis), and fringes appear above and below the line; the greater the load, the greater the number of fringes. From the patterns produced, it is difficult to determine which fringes indicate tensile forces and which indicate compressive forces, so it is suggested that reference is made to the previous rubber-model demonstration.

Strain-gauge techniques

iii) Equipment required (see *Appendix 2*): Perspex angle 12 × 12 × 3 mm with strain-gauge attached.

Using the Perspex angle with the strain-gauge attached, insert the 4 mm plugs and zero the instrument as described in *Appendix 2*.

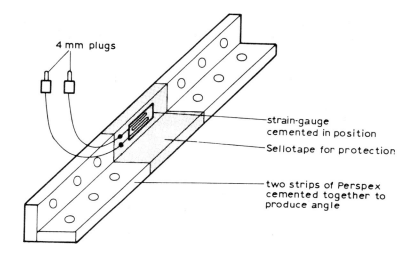

4 mm plugs

strain-gauge
cemented in position

Sellotape for protection

two strips of Perspex
cemented together to
produce angle

Tension the angle by pulling on each end. Notice in which direction the pointer moves.

Next place the angle vertically, and press on the end to produce compression. The pointer now moves in the opposite direction.

Further demonstrations could show that the internal forces in the member might change if the external forces were applied in different positions, e.g., a cantilever.

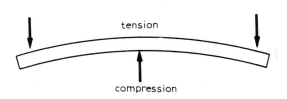

tension

compression

The reference made in the follow-up material to the relative influence of the depth and breadth of beams should not be extended to include the second moment of area calculations in beam-bending stress theory. Such reference is outside the concepts of this course.

However, pupils' attention can be drawn to visual examples where the concept has been applied, such as castellated beams and open-lattice structure beams.

a castellated beam

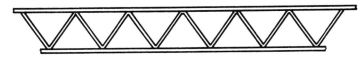

an open-lattice structure beam

2 Demonstration to indicte tension or compression in individual members of a structure

Pupils were asked to indicate, on the sketches of their final design, the forces acting in individual members of the structure.

It is fairly easy to do this for certain members, but more difficult for others. If the load is moved, the forces acting in particular members may change from tensile to compressive and vice versa.

To indicate clearly whether a member is in tension or compression, strain-gauging techniques can be used. The Meccano member under investigation would be removed and replaced with the Perspex angle girder to which is attached a strain-gauge.

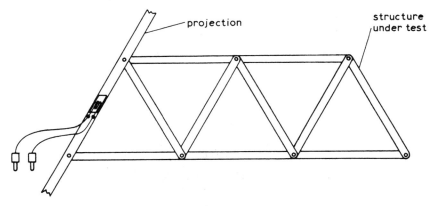

projection

structure under test

The strain-gauge is connected to the instrument, the meter is adjusted to zero, and the load is then applied. The direction of movement of the pointer indicates whether the member is in tension or compression.

The meter indicates not only the type of force, but also the magnitude of the force.

Buffer Experiments

Use two spring-balances to support the bridge. From this the pupils should discover that:

    the total upward force = the total downward force

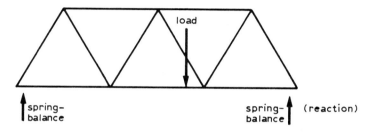

A teacher demonstration could then be given to illustrate the principle of moments by moving the load along the bridge.

Homework Questions

1   If the child has a mass of 30 kg what is the tension in each rope? (Neglect the mass of the seat and ropes.)

2   If two boys, A and B, carry a mass of 100 kg placed upon a
    pole, as shown, find the proportion of the mass carried by each
    boy. (Neglect the mass of the pole.)

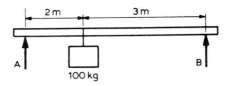

3   Indicate with a 'C' those members in compression and with a
    'T' those members in tension in figs (a) to (e) below. If a
    number is in bending, indicate this also, showing which side is
    in compression and which in tension.

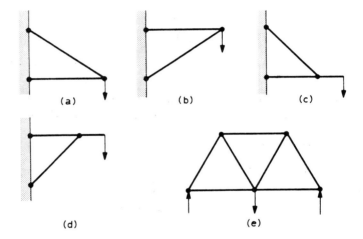

Structures Assignment 4

A further mini-project is now suggested in which pupils are
expected to apply principles which they have already learned and to
consider the problem of stability of free-standing devices. A specification is given for a lifting device for loading vehicles. Again we
emphasise the importance of design on paper before construction
takes place. We now bring in consideration of the cost and weight
of such a structure by asking that the final design shall be as stable

and as rigid as possible without incorporating unnecessary members.

Most pupils realise that a member under tension can be replaced by wire supports, and some interesting devices can be produced using a number of wire supports in place of flat strip. Interesting discussion points also arise as to the possibility of constructing a crane completely from wire (i.e. all members under tension). As the teacher goes round the various groups, he should ask questions such as, 'Is that member under tension or compression?', or, 'Have you included any members which are unnecessary?'

An assortment of lengths of wire with bolts at each end for securing them rigidly to the structure could be made available. Cut a number of different lengths of wire of about 26 s.w.g. (0.45 mm diameter). Drill a hole in the head of a Meccano bolt, as shown, thread the wire through the hole, and twist.

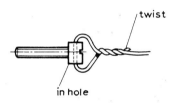

The teacher may feel it wise to wait until definite lengths are requested, and to have long lengths of wire and a supply of ready-drilled bolts available. The ends should be well twisted to prevent slipping.

As each model is completed, the teacher should comment on the result, perhaps offering further suggestions. These suggestions, however, should be carefully worded. For instance we might say, 'Are you sure that this member should be made from flat strip?', but never, 'This member is under tension so it must be made from flat strip rather than angle, to reduce the weight of the crane.' In short, questions should be phrased in such a way that the answers come from the pupils. Pupils must be confident that they know why their device is made the way it is. Later, when the pupils start major projects, it is all too easy for the teacher to provide solutions to the problems which arise.

Only when the model has been satisfactorily completed should reference be made to the follow-up sheet.

Apparatus Required (15 pupils, 5 groups of 3)

assorted Meccano angle girder and strip, flanged plates, etc
five lengths of cord and hooks
five electric motors
assorted gears
five 1 kg weights
assorted pulleys for cord
short axles

Some teachers may wish to have a model vehicle available.

Demonstrations

The work should be followed by demonstrations to consolidate the understanding of tension, compression, bending, and torsion. Demonstrations including photoelastic stress analysis and strain-gauge techniques, if not already covered, could now be given.

If necessary, demonstrate the effect of moments by placing weights at various distances from the pivot of a see-saw arrangement.

Buffer Experiments

Whenever pupils are attempting assignments involving the design and construction of a device, the faster groups can be extended by encouraging them to make their device more reliable or sophisticated.

Homework Questions

1   What vertical force will be required at X to prevent the crane from overturning?

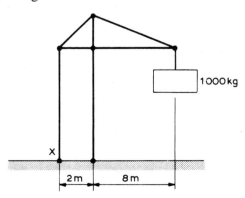

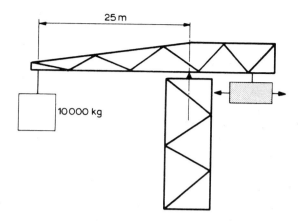

2   At what distance from the axis of rotation should the centre of the balancing mass be placed? (Neglect the mass of the structure.)

3   Analyse the forces acting in the members of this structure and indicate with 'C' and 'T' whether they are in compression or tension.

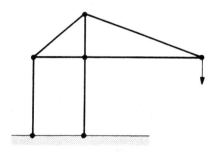

Gears

The work already covered in the *Structures* section results in the production of more or less static mini-projects. The *Gears assignments* are concerned with motion and the transmission of mechanical energy. Many pupils are fascinated by devices which incorporate moving parts, where something can be seen to be happening. *The Gears assignments* and the *Basic electricity assignments* form a link between static devices and the design, construction, and use of devices involving motion.

In *assignments 1*, pupils examine the properties of a number of common gear systems. Pupils' diagrams of spur-gear systems need consist only of circles in contact; the number of teeth on a particular gear being written within the appropriate circle (e.g. 57T).

Assignment 2 introduces chain and belt drives. Pupils are asked to compare these two systems with spur-gear systems.

Concepts of friction and mechanical efficiency are studied in *Gears assignments 3* and *4*. The former is illustrated by experiencing the effort required to turn certain gear shafts, whilst the latter is covered by making use of the multi-speed electric motor and gearbox. This versatile motor is used frequently in the course. An investigation into its capabilities is, therefore, appropriate at this stage.

Gears assignment 4 includes the use of electrical measuring instruments: the ammeter and the voltmeter. Unless the pupils have some prior knowledge and experience in the use of meters, this assignment should be deferred until after the *Basic electricity assignments. Gears follow-up 4* may need further explanation for less-able pupils.

Gears Assignment 1

For this programme, it is suggested that the class is divided into five groups of three pupils, each group investigating, in turn, the five different gear systems provided. These five systems demonstrate various ways of transmitting rotary motion and should be constructed before the commencement of the lesson. Some construction notes are included, but these are intended only to give guidance. Other arrangements may be found more satisfactorily, or different constructional material could be used, e.g. the mounting frame could be made from Perspex, the axle holes being positioned by using a Meccano strip as a template.

If the investigations are carried out carefully, and a full record is made, pupils should experience little difficulty in drawing conclusions which apply to all gear systems.

The Meccano range includes bevel and helical gears and, if these are available, it is suggested that units using them are constructed and shown to the class. However, these gears are expensive and more difficult to use than the contrate and spur gears.

Apparatus Required

Various pairs of Meccano gears can be selected to give whole number ratios,

e.g. 25T and 50T
      19T and 38T or 57T or 95T or 133T
      15T and 60T

In the programme, particular gears are specified to simplify the follow-up sheets; if these gears are not available, others may be used provided the notes are modified accordingly.

With the simple system, little support is needed for the gear shafts, but if improved bearings are required, as for example in System 3, it is suggested that part number 62b (the double-arm crank) be used for this purpose.

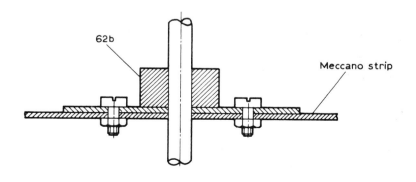

Construction of gear systems

System 1: simple train of spur gears.

Parts required	Part no.
one 19T gear	26
one 57T gear	27a
two axle rods, 50–75 mm	
one flanged plate, 60 × 38 mm	51
two collars	59
one washer	38

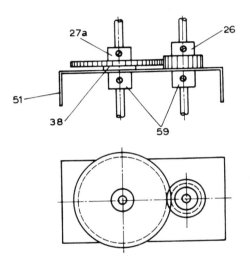

System 2: simple train of spur gears with an idler

Parts required	Part no.
one flanged plate, 90 × 60 mm	53
two 57T gears	27a
one 19T gear	26
three collars	59
two washers	38
two axle rods, 50–75 mm	
one axle rod, 25 mm	

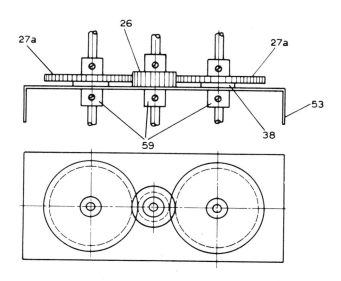

System 3: compound train of spur gears

Parts required	Part no.
one flanged plated, 90 × 60 mm	53
two 60T gears	27d
two 15T gears	26c
three axle rods, 50–75 mm	
three double-arm cranks	62b
1 washer	38
four collars	59

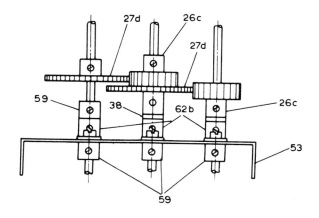

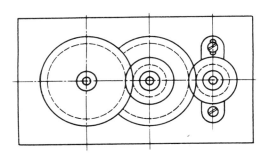

System 4: spur gear and contrate gear

Parts required	Part no.
one flanged plate, 50–75 mm	51
three trunnions	126
one double-arm crank	62b
three collars	59
one 50T contrate gear	28
one 25T pinion	25
one axle rod, 100 mm	
one axle rod, 50 mm	
one washer	38

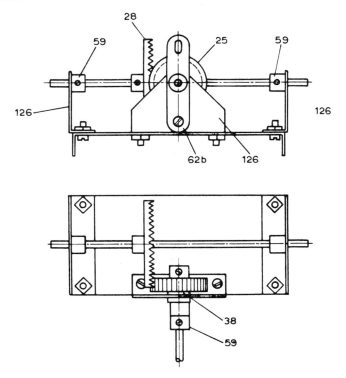

System 5: worm gear and pinion

Parts required	*Part no.*
one flanged plate, 50 × 75 mm	51
four trunnions	216
one 19T pinion	26
one worm	32
four collars	59
one axle rod, 100 mm	
one axle rod, 75 mm	
eight washers	38

(washers to be used as spacers *under* the trunnions.)

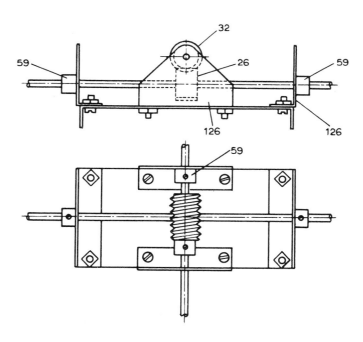

Demonstrations

1 If Meccano bevel and helical gears are available, these systems should be demonstrated.

2 Gear systems in everyday use can be demonstrated in the following applications:
   spur gears (lathe screwcutting change gears; camshaft drive on some engines);

worm and pinion (tennis-net tensioner);
bevel gears (car rear-axle differential);
helical gears (car distributor drive from camshaft; constant-mesh gearbox).

These demonstrations should lead to a general discussion on the use of gear systems.

The following points might arise:

quieter operation — advantages of helical gears;
precision shaping of gears;
critical positioning of gear shafts;
lubrication;
advantages of worm-and pinion drives.

Buffer Experiments

1   If a car gearbox or back axle is available, pupils should determine the ratios between input and output shafts. Similarly, lathe screwcutting gear trains could be investigated. (Ensure the lathe is electrically isolated before the guard is removed.)

2   Devise a method of measuring the approximate forces needed to turn the input and output shafts.

Homework Questions

1   How many times must gear A be turned for gear B to turn once in the gear system shown below?

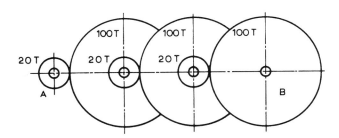

2   A car engine drives a shaft to which is fixed a gear with 20 teeth. A reduction in shaft speed of 3.5:1 is required (i.e. the engine turns 3.5 times to turn the gearbox output shaft once). How many teeth are required on the gearbox shaft gear?

Gears A and B rotate on shafts in fixed positions. Gear A can be rotated only in the direction shown; B, however, must have provision to be driven in either direction. Can you suggest a solution to this problem?

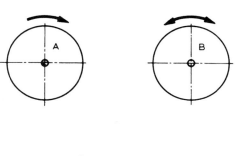

The diagram shows a rack and pinion (ensure that you have seen one of these before attempting the question). If the pinion rotates at 50 rev/min, how fast does the rack travel, and in which direction?

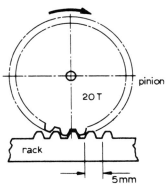

The motor in a electric clock runs at 300 rev/min. What gear ratios are required to drive (a) the second hand, (b) the minute hand, (c) the hour hand of the clock?

Gears Assignment 2

This programme deals with the transmission of rotary motion by chain and belt systems. The investigations, again carried out in small groups, are similar to those in *Gears 1*. Is is suggested that two of each type of system are constructed, since only three models are used.

After examining various ways of transmitting rotary motion, pupils should have gained sufficient knowledge to:

i) select the most suitable transmission system for a given purpose;

ii) calculate the sizes of gears, pulleys, etc. required in a system.

53

This will enable them to solve, with confidence, the transmission problems that they are likely to meet in future practical examples.

It is suggested that some classroom time be devoted to working through a few typical problems, so that pupils are able to tackle them in a methodical way.

Apparatus Required

Two of each of the following systems should be made; six units in all. All systems are constructed in a similar way.

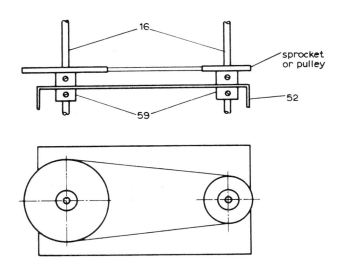

System 1: Chain-and-sprocket system

Parts required	Part no.
one flanged plate, 140 × 60 mm	52
one 25 mm, 18T sprocket	96
one 50 mm, 36T sprocket	95
two axle rods, 50 mm	
two collars	59
one chain	94

Systems 2 and 3: Pulley-and-belt system

Parts required	Part no.
one flanged plate, 140 × 60 mm	52
one pulley, 25 mm dia.	22

one pulley, 50 mm dia.	20a
two axle rods, 50 mm	
two collars	59
one driving band	186c
or spring cord and	58
coupling screw	58a

The bearings for the shafts may be improved by using two 62b double-arm crank as described in *Gears 1*.

Demonstrations

1 A bicycle chain-and-sprocket drive mechanism should be examined. The construction of the chain could be discussed. How is the chain tensioned and why? Why not use a belt or spur gears?

2 Various pulley systems should be examined in which flat, round, and vee-type belts are used,
    e.g. bench drilling-machine or lathe drive (vee belt),
        car engine-fan and generator drive (vee belt),
        sewing machine, either treadle or electrical (round belt),
    some workshop milling machines or older machines may use flat belts.

Discuss the advantage of the vee belt over the flat belt. Stress the importance of tensioning the belt, and discuss the various ways of doing this.

3 If a car engine is available, discuss the method of driving the camshaft from the crankshaft, and compare this with the fan and generator drive. Why is a chain normally used for one drive and a belt for the other?

This is an ideal time to discuss the new cogged belt used for driving some overhead camshafts.

Buffer Experiment

Calculate the overall gearing of a bicycle, i.e. the distance the bicycle moves forward for one revolution of the pedals. Pupils should then verify their answer by moving the pedals exactly one revolution and measuring the distance the bicycle moves forward.

Homework Questions

1   Make sketches of the following two diagrams:

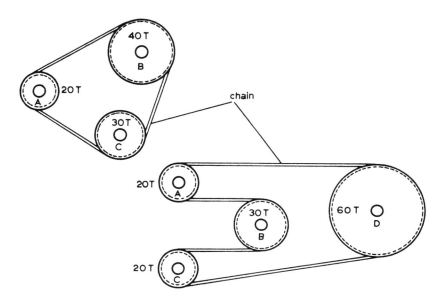

indicate on each
i)   the direction of rotation of each sprocket, and
ii) the number of revolutions each gear makes if chain wheel A is turned three times.

2   What is the purpose of pulley A in the system shown below?

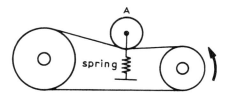

3   A bicycle pedal chain wheel has 48 teeth, and the rear sprocket has 18 teeth. If the pedal is rotated one complete revolution, how many times will the rear wheel rotate?

4   An electric motor fitted with a 100 mm diameter pulley revolve at 1500 rev/min. It is required to drive a circular saw at

2500 rev/min. What size of pulley must be fitted to the saw spindle?

5 A bicycle has 560 mm dia. wheels (including the tyres) and a pedal chain sprocket with 48 teeth. Three different sprockets are to be fitted to the rear wheel hub to provide three 'gears'. For each revolution of the front chain sprocket the bicycle should move distances of (a) 3.52 m, (b) 5.28 m, (c) 7.04 m. How many teeth are needed on each of the three rear sprockets?

Gears Assignment 3

With pupils of above-average ability, this programme could be carried out by means of group work, provided that five sets of apparatus are available. However, with pupils of average and below-average ability it is recommended that one set of apparatus be used and that the teacher controls the experiments. Questioning should ensure that the pupils understand the principles involved.

A certain amount of additional teaching may be necessary, depending on the ability range and background knowledge of the pupils. For instance, considerable time could be spent on defining the terms 'work' and 'energy'. The mathematics involved should be kept simple, and perhaps left out altogether for the less able. In this event, *Gears assignments 3* can be attempted by the pupils, but *Gears follow-up 3* should be replaced by a simplified version.

Apparatus Required

System 1

Parts required	Part no.
Two flanged plates, 140 × 60 mm	52
Two angle girders, 320 mm	1
Four strips, 75 mm	4
One 50T gear	27
One 25T gear	25
Two axle rods, 75 mm	
Two anchoring springs	176
One hank of cord	40
One 1 kg mass	
One spring-balance, reading to 1 kg	

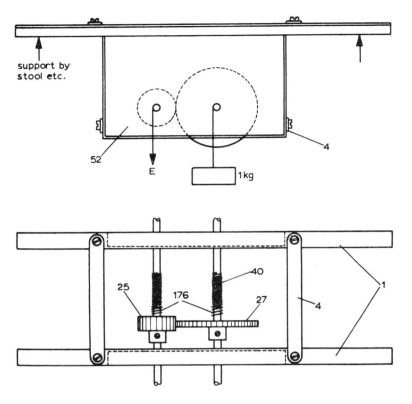

System 2: Nylon mounted axles

The list of parts required and the construction is identical to that in System 1, except for the two axle rods. In addition, four nylon bearings and eight 4 BA nuts are required, assembled as shown.

Replace two 75 mm axles with two axles as shown below.

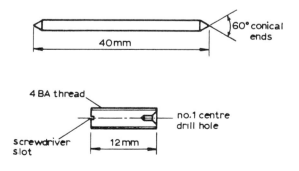

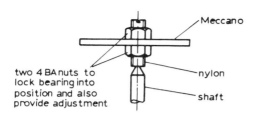

If nylon is not available, a similar, well-lubricated brass bearing would provide greater efficiency than that provided by System 1.

Demonstrations

1. Illustrate the gear systems incorporated in overhead cranes etc.

2. Simple concepts of work and energy could be illustrated in the following way:

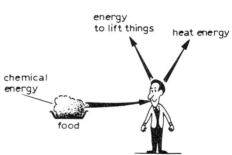

i) Discuss the meaning of energy in general terms. E.g. we eat food containing chemical energy, and within the body this energy is converted into

   a) heat energy, and
   b) energy required to lift things or to perform other work.

ii) Lift 1 kg a vertical distance of 1 metre (say from the floor to a bench or low shelf). Establish the fact that energy from food has been used. Where has the energy gone?

Allow the mass to fall to the floor and to break safely a piece of wood etc. Establish that the energy was 'given' to the mass and that, when this fell, some energy was used in breaking the wood

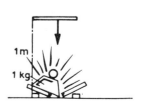

59

and the remainder was lost as heat. (Mention the heat generated when a hammer is used on metal.)

iii) Lift 1 kg a vertical distance of 2 m. How much energy has been 'given' from chemical energy, and how much has the mass been 'given' in its new position?

iv) Lift two 1 kg masses (or preferably one 2 kg mass) a vertical distance of 1 m, 2 m, etc. Establish that the energy transferred to the mass depends upon the vertical height raised and the mass.

Discuss the fact that 1 kg lifted diagonally to a vertical height of 1 m will do no more 'damage' in falling that 1 kg lifted vertically, i.e. the energy given to the mass is the same.

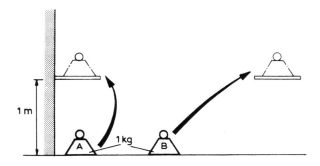

Establish using numerical values at first)
   energy given to an object = force × distance (vertical)

Discuss units: N (force in newtons) × m (distance in metres)

N.B. The gravitational force acting on a mass of 1 kg is about 10 newtons.

Hence energy given to an object = Nm (newtons metres).

One newton metre (Nm) is more commonly referred to as one joule (J).

v) Discuss the term 'work' as a measure of the amount of energy transferred from one form to another.

e.g. when 1 kg (10 N) is lifted a vertical distance of 2 m, position (potential) energy given to the 1 kg mass
= force × distance (vertical)
= 10 × 2 Nm

$$= 20 \text{ Nm}$$
$$= 20 \text{ J}$$
∴ amount of potential energy transferred = 20 J
∴ work done in lifting the 1 kg mass = 20 J

vi) Several other simple calculations should be carried out.

Buffer Experiment

1 i) Repeat the assignment using a worm and pinion system. Calculate its efficiency and compare it with the previous systems used.

ii) Thoroughly lubricate the worm gear and pinion and the shaft bearings. Is there an increase in efficiency?

2 Fix a small pulley to the shaft of the small gear. Attach the cord to the pulley, instead of to the shaft, and repeat the experiment. How do the results compare with those obtained in the assignment? Try to account for any differences.

3 Attach the 1 kg load to the cord on the small gear shaft and apply the effort to the cord on the large gear shaft. Compare your results with those in the assignment.

Homework Questions

1 If a simple two-gear system has a velocity ratio of, say, six and one gear has fifteen teeth, how many teeth has the other gear?

2 A 100 kg load attached by a cord to the shaft of a 40T gear wheel driven by a worm is just raised by an effort of 100 N acting on a cord around the other shaft. What is the efficiency of the system if both shafts are of the same diameter?

3 In a similar gear system incorporating a 100T gear driven by a 25T gear, an effort of 2500 N is required to steadily raise a load of 600 kg. After the gears and shaft bearings have been well lubricated, a 2000 N will just raise the same load. What is the increase in efficiency of the system?

4 A machine raises a load of 180 kg a distance of 2 m. The effort, a force of 240 N, moves 18 m during the process. Calculate the velocity ratio, mechanical advantage, and the efficiency at this load.

Gears Assignment 4

As this investigation involves the use of both a voltmeter and an ammeter, it is recommended that the *Basic electricity* section is completed before this programme is attempted.

Although pupils find the experiment well within their capabilities, care must be taken to switch off the motor in good time, before the load carrier reaches the Meccano frame. If fouling does occur, the motor gearbox will probably be damaged. For this reason, the use of a gear ratio below 6:1 is not recommended, because the load is then lifted too quickly.

It is recommended that pupils measure the time taken for the load to rise 0.5 m, as this time is referred to in the *follow-up sheets*. This investigation can be developed further depending on the ability and background knowledge of the pupils. Using a variety of different gear ratios the 'work done' in each case can be calculated. Gear ratios up to 360:1 can be investigated and even higher ratios can be reached by including extra gear units. But it must be remembered that at very high ratios the gearbox does not have the mechanical strength needed to lift the heavy 'theoretical' loads. Therefore gear ratios above, say, 60:1 should be regarded as useful for reducing the rotation speed of the output shaft but not as a means of increasing the available torque.

If the pupils have a sound knowledge of electricity, it might be possible to determine the overall efficiency of the system by calculating the electrical energy used by the motor and the work done by the motor in lifting the load.

Apparatus Required

Lifting device (five required)

Parts required	Part no.
two angle girders, 320 mm	1
one angle girder, 240 mm	1a
two flanged plates, 90 × 60 mm	53
two flanged plates, 60 × 38 mm	51
two strips, 60 mm	5
four corner brackets	133
one axle rod, 130 mm	
one flexible coupling	175

two couplings	63
one anchoring spring	176
one hank of cord	40
one multispeed electric motor	

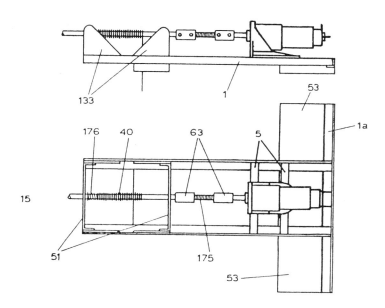

Other items required:
ten weights to stand on platform to provide stability (two per group);
ten loads (1 kg) consisting of ten 100 g masses (two loads per group);
five loads (100 g) consisting of ten 10 g masses (one load per group);
voltmeters reading 0–10V;
ammeters reading 0–1A;
12 V d.c. power supplies;
leads with 4 mm plugs;
a stop watch or other means of measuring time intervals of a few seconds.

The original Meccano motors have now been replaced with the much sturdier multi-speed units. These have one drawback in that it is no longer such a simple task to change speed, as four screws must now be removed and the gearbox reassembled (See Equipment Guide for details).

It is recommended, therefore, that a set of five or six lifting device units be made up, each with a different gear ratio, (clearly marked) and the class use these on a 'circus' basis.

In this case, as each group will be starting with apparatus provided with a different gear ratio, it is important that they each use an appropriate load for their first experiment. It is necessary, therefore, to tell each group what load to use first. Experience has shown that the following loads are suitable, when operated on a 6 volt supply.

Ratio	Load lifted
6:1	200 grams
12:1	400 grams
18:1	600 grams
24:1	800 grams
30:1	1000 grams

It is important not to 'mark' the above loads on the lifting device, as pupils would obviously be aware of expected results on their subsequent tests:

NOTE: Experience has shown that the assembly of the lifting device without the use of the flexible coupling (part no. 175) produces poor results. (A rigid coupling of slightly mis-aligned shafts causes the current flow through the motor to vary during each revolution of the output shaft.)

The pupils could change ratios themselves but this would involve removing the motor, dismantling and reassembling the gearbox and remounting the motor. This would take about 10 minutes.

Demonstrations

1 A possible demonstration for the more able pupils is suggested under 'Buffer experiments'.

2 Demonstrate the effect on the current consumption of the motor, and on lifting speed, when the layers of cord pile one on top of another. Establish that the effect is similar to changing the motor to a lower gear ratio.

3 Demonstrate the 3:1 ratio to show that the lifting speed is

excessive, warning that damage can occur to the motor if it is not stopped in time. Pupils are instructed not to use ratios below 6:1

4   Adjust the motor to engage the 6:1 ratio. Lift a 1 kg load and note the voltage and current. Switch off the supply (holding the load) and show that, when the load is released, electrical energy is generated as indicated by the voltmeter. This emphasises that the electrical energy produces mechanical (potential) energy which can be used to produce electrical energy again. Light a 12 V, 2.2 W bulb, by placing a lamp indicator unit in the circuit, in addition to showing a meter reading.
(Note: To obtain positive voltmeter readings when the motor acts as a generator, the reversing switch on the motor must be changed over.)

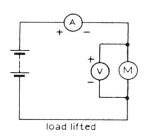

load lifted

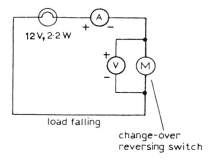

load falling

change-over
reversing switch

Buffer Experiments

1   Repeat the assignment, this time measuring the vertical distance that the load is lifted in some specific time.

Calculate and compare the work done by the motor when alternative gear ratios are engaged. (i.e. compare the work done in a given time — power or rate of working).

2   The overall efficiency of the system can be investigated by the whole group (if they are of above average ability) or covered by demonstration. Set up the apparatus as before (*assignment 4*). Choose a gear ratio and a suitable load (F newtons)

Record:
i)   the time (*t*) in seconds taken for the load to be lifted through a given distance (*l*) in metres;

ii) the ammeter reading (*I*);

iii) the voltmeter reading (*V*).

Calculation:

Rate of consumption of electrical energy (i.e. power supplied to motor) = VI joules/second

∴ energy supplied to the motor (in time *t*) = VIt joules (or newton metres)

Work carried out by the machine (in time *t*) = force × distance (vertical) = $F \times l$

$$\text{Efficiency} = \frac{\text{useful work carried out by machine}}{\text{total energy put into machine}}$$

$$= \frac{F \times l}{VIt} \times 100\%$$

Homework Questions

1 Explain why it is virtually impossible for a motor car to move off, from rest, in top gear.

2 The diagram shows a device for lifting water from a well. If the bucket and water exert a force of 200 N what is the least possible force needed to turn the handle? If the rope coils were to pile on top of each other, what effect would this have on the force needed to turn the handle?

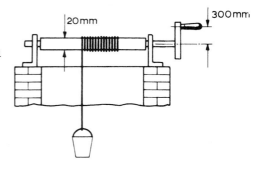

3 The gears in the gearbox of the electric motor you have used are most easily damaged through overloading when a high ratio, say 60:1, is used. Explain why this is so.

Basic electricity

This section of the course is included to give pupils sufficient background knowledge of electricity to enable them to tackle the *Electrical switching assignments* which follow. They should have a working knowledge of current, voltage, resistance, the use of the voltmeter and ammeter, and be familiar with common electrical symbols. Three programmes are included, though some teachers may wish to omit *Basic electricity assignment 3*, which deals with Ohm's Law. During the second year of the course, applications of Ohm's Law are dealt with more fully in the *Electronics: resistance assignments*.

Some teachers may be tempted to devise further programmes for this section of the course, but it is recommended that they should first cover the complete course at least once, in order to fully appreciate the purpose of the *Basic electricity* programmes and the manner in which they link with others.

Basic Electricity Assignment 1

Firstly a number of basic electrical symbols are introduced. Perhaps the most important of these, at this stage, is the switch symbol. Emphasis should be placed on the use of a 'solid' circle to represent the pole of a switch and the 'open' circle to represent contacts. Later programmes include fairly complex switching-circuit diagrams and, unless the poles are clearly marked, the circuits are difficult to follow.

The pupils are asked to connect a single bulb to a single cell and a switch, and to observe the effect when the switch is closed. For this experiment a 1.5 V cell should be used in conjunction with a 1.25 V lamp.

Having seen the bulb illuminated with the cell one way round, the pupils are then asked to reverse the cell. An ammeter is then used (1 A full-scale deflection, approximately) to find the current taken by a single lamp run at its normal intensity. When two, then three, cells are connected in series with one lamp, it is advisable to suggest that the tests should be as short in duration as possible, otherwise bulb filaments are likely to 'blow'. These experiments are important since they give the pupils an insight into the relationship between voltage and current and the effect of fitting cells in series.

Parallel circuits are investigated using lamps connected in parallel. Pupils should learn from this that currents can divide and that currents can be measured in a circuit by fitting an ammeter anywhere in it. In effect they simply 'cut' a lead and insert the meter. Should the resistance of the meter cause a lamp brightness to diminish, then a fruitful discussion on the 'resistance' of ammeters could follow. The currents in a parallel circuit are compared with those in a series circuit. Brighter pupils may suggest that the lower brightness of series bulbs is due to the increased circuit resistance.

Apparatus Required (15 pupils, 5 groups of 3)

cells: three per group (1.5 V) in a carrier unit fitted with 4 mm output sockets.

lamps: four per group (1.25 V, 0.25 A)

switches: one per group, single pole on-off. The double-pole, double-throw toggle switch unit, or the single-pole, double-throw push-button switch could be used

ammeters: one per group, approximately 1 ampere full-scale deflection

leads: nine per group (fitted with stackable 4 mm plugs)

Demonstrations

A number of different types of cell, and ammeters with different scale deflections, could be demonstrated.

Buffer Experiments

1   If enough lamps and holders are available, try more than two lamps in series and in parallel.

2   Offer a universal meter, to replace the ammeter in the circuits, and try the effect of using different ranges. Ask the pupils to decide whether a large or small deflection of the pointer gives the most accurate value for the currents flowing.

3   Provide a relay, and ask the pupils to discover at what current the relay energises.

4   Provide a high-current solenoid. Ask the pupils to connect it
    across eight U2-type cells in series, and compare the results with
    the effect when the solenoid is connected across a car battery.

Homework Questions

1   Draw a circuit diagram to show how you would connect two
    6 V lamps to a battery giving 12 V so that each will light
    normally. How much current would you expect to be drawn
    from the battery, compared with that drawn when one of the
    lamps is used?

2   Draw a circuit diagram to show how you would connect two
    12 V motors in series with a 12 V battery.
    What can you say about the speed of each motor, compared
    with the speed of a single 12 V motor connected to a 12 V
    battery?

3   Would you expect motor-car lamps to be connected in series or
    parallel? Show how you would connect an ammeter into a
    motorcar circuit in such a way that it would register the current
    taken by any combination of lamps.

4   Draw a circuit diagram containing one on-off switch, two
    ammeters, two 6 V lamps, two motors, and a 6 V battery. The
    lamps should glow less brightly than normally. One ammeter
    should read the current taken by one motor and the other the
    total current in the circuit. The switch in its 'off' position
    should prevent any current being taken from the battery.

Basic Electricity Assignment 2

This programme introduces the voltmeter as a means of measuring
electrical 'force'. Cells can be fitted in parallel using a cell carrier.
Pupils will discover, at a later date, that not all power supplies are
capable of providing large currents.

Basic electricity follow-up 2 raises the question of the resistance of
a circuit, and it is intended that at this stage the teacher should
discuss resistance in more general terms with the pupils. Should any
ideas be put forward, then the class could devise experiments to test
them.

Apparatus Required (15 pupils, 5 groups of 3)

cells: two per group (1.5 V), in suitable carrier

lamps: one per group

switches: one per group, single-pole on-off

ammeters, one per group, approximately 1 ampere full-scale deflection

voltmeters: one per group, approximately 5 volts full-scale deflection

accumulator: 12 V motor-car lead-acid battery

Demonstration

1 Allow pupils to see and examine a number of voltmeters with different ranges. Demonstrate a universal meter such as the Avometer. If appropriate, mention alternating current, and also that the Avometer is one of the types of meter which will measure both a.c. and d.c.

2 Demonstrate the low internal resistance of a motor-car battery — ensure it is well charged — by shorting out the negative and positive terminals with one strand of a length of multi-stranded insulated wire. Repeat the procedure using eight U2-type cells in series, and compare the results.

Buffer Experiments

1 Use a universal meter to replace the voltmeter and ammeter.

2 Replace the lamps with an electric motor and measure the current flowing in the motor with differing supply voltages. Find a relationship between motor speed and supply voltage.

3 Investigate the current carried by the motor under differing load conditions. The motor can be loaded by trying to stop the motor shaft turning by gripping with the fingers — this is not recommended on ratios above 30:1, as damage to the gearbox may result.

Homework Questions

1 If you were Chief Engineer at a power station feeding only one small town, how would you prevent damage to the generators if

70

the current consumed by customers was increasing to a dangerous level? Suggest *three* different methods.

2   Two 6 V lamps with a voltmeter across each are connected in series with a 12 V battery. Each lamp glows with the same brightness. Suddenly one lamp becomes very dim and the reading on its voltmeter falls. The other lamp immediately becomes very bright and its voltmeter reading increases. Suggest what could have happened to give this effect.

3   A machine requires a 12 V supply, and it consumes a large current. You have eighteen cells of 2 V. Show how you would connect the cells to the machine so that it would run normally for the longest time possible.

Basic Electricity Assignment 3

This programme could be attempted by the brighter pupils, who will benefit by knowing the relationship between voltage, current, and resistance. Then the knowledge should be consolidated by calculating relay currents and resistances when relays are used in switching circuits.

Apparatus Required (15 pupils, 5 groups of 3)

Variable voltage supplies, such as the Universal Labpack produced by Radford Laboratory Instruments Ltd: one per group. This supply will provide an a.c. or d.c. output from 0 to 25 V at 8 A maximum in 0.2 V steps.

Resistance wire: a reel of resistance wire such as Eureka 26 s.w.g. (0.45 mm diameter).

Demonstrations

If variable-voltage supply units are in short supply, some teachers may wish to demonstrate this programme.

Buffer Experiments

1   Graph the results of one of the experiments, plotting voltage against current. Repeat the experiment with a filament lamp, to demonstrate the increase in resistance of the filament at a high temperature.

2   Graph the voltage/current relationship for a semiconductor diode.

Homework Questions

1   If in the circuit shown below the voltmeter reads 6 V and the ammeter 2 A, what is the effective motor resistance?

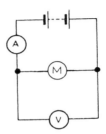

2   The heating element in a small electric furnace has a resistance of 50 ohms. What current flows in the element when it is connected to a 250 V supply?

3   The filament in a lamp is most likely to 'burn out' when it is first switched on. Why do you think this is so?

4   If in the circuit shown below A_1 reads 1 ampere, A_2 reads 2 amperes, A_3 reads 3 amperes, and the voltmeter reads 6 V,

   i)   which lamp will give out the most light?
   ii)   which lamp has the lowest filament resistance?
   iii) what current is taken from the battery?
   iv)   what is the resistance of the filament of L_1?
   v)   what is the resistance of the filament of L_2?
   vi)   what is the resistance of the filament of L_3?

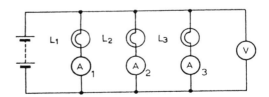

Electrical switching

Pupils attempting this section of the course should have some knowledge of the fundamental principles of electricity. It is important that they have been introduced to basic ideas of voltage, current, and resistance and have measured voltages and currents. They should also have a knowledge of the common electrical symbols, such as those for the cell, battery, resistor, switch, and lamp, and should be able to draw simple circuits. Ideally, the *Basic electricity assignments*, which cover these points will have been attempted before beginning the *Electrical switching assignments*.

The electrical symbols used throughout the textbooks are those recommended in B.S.I. publications P.D. 7307: 1982. which are based on the symbols used in B.S. 3939. Some variations may arise with pupils who use symbols seen in textbooks and magazines which are based on B.S.I. 1852. The symbols for resistors vary; e.g., 4K7 for 4.7K Ω, 390R for 390 Ω and 1M2 for 1.2 MΩ. However the differences are not very great and pupils should have little difficulty in translating them.

The *Electrical switching assignments* are intended to introduce pupils to methods of electrical control and to some of the more important sensing devices used in control systems. The electromagnetic relay is used frequently, and pupils become familiar with relatively complex switching circuitry involving delays, flip-flops (bistables), and clocks (oscillators). The operation of a relay is easily understood by the average pupil. Experience has shown that many pupils who have only a modest knowledge of magnetism and electricity can reason how the device operates, following brief examination. What is not so easily grasped is the relationship between the applied voltage, the resistance, the current, and the number of turns on the coil for a given relay. When pupils have used a number of different relays in a variety of situations, they will more easily appreciate these factors and consider them when designing circuitry for major project work later in the course. Whenever appropriate, therefore, we should encourage pupils to think about relay circuitry design in terms of the relay they are using at a particular time.

The *Electrical switching assignments* can be summarised as follows:

Assignment 1: A vehicle is designed to meet a given specification.

Here the pupils apply their knowledge of structures and gearing to produce a rigid vehicle moving at a specific speed.

Assignment 2: The use of a two-pole, two-way switch unit is investigated by pupils to discover which pairs of sockets are connected in the two possible positions of the switch toggle. Pupils use switch units to stop, start, and reverse the vehicle.

Assignment 3: Pupils use reed-switch and microswitch sensors to indicate when the vehicle strikes an obstruction.

Assignment 4: The relay is introduced through the two-pole, two-way relay unit. Pupils investigate the switching in a similar manner to that used in the switch-unit investigations. They are asked to design and assemble a circuit consisting of a reed-switch and relay to make the vehicle reverse when it strikes an obstruction.

Assignment 5: Using a microswitch and relay, pupils set up a circuit which will enable their vehicle to reverse when it strikes an obstruction. A capacitor is fitted across the relay coil to produce a time delay before the relay de-energises. Using a number of different capacitor values, pupils discover an approximate relationship between delay time and capacitance.

Assignment 6: Here the bistable principle is introduced. Pupils investigate a relay-bistable unit and then use it to cause their vehicle to reverse when it strikes an obstruction, and to continue reversing until a second obstruction is encountered.

Assignment 7: Pupils now investigate the use of the light-dependent resistor (photocell). Using the cell and relay, pupils are asked to make their vehicle stop when it breaks a beam of light.

Assignment 8: Using a photocell and one relay, pupils are asked to design and test a circuit which will cause their vehicle to oscillate in and out of a light beam. Three alternative values of capacitor are fitted across the relay coil, and pupils observe the effect in each case.

Assignment 9: Pupils again attempt to make their vehicle reverse out of a beam of light, but a second relay is incorporated. The operation of this second relay is delayed.

Assignment 10: Using a relay-bistable unit and two photocell units, a vehicle is made to oscillate between two beams of light.

NOTE: It is recommended that *assignments 9* and *10* become teacher demonstrations for all but the most able children. Some teachers may also wish to demonstrate *assignment 8*.

For the first *Electrical switching assignments*, it is perhaps advisable to allow pupils to record their circuits simply as switch-unit diagrams. As soon as it is expedient, however, encourage the drawing of electrical circuit diagrams alongside, and eventually discourage the use of any type of diagram which is not conventional.

British Standards suggest the following means of showing wires which are electrically connected and those which are not.

A, B, C connected   A and B not connected

The following convention is sometimes used, but is now out of date and should be discontinued.

C   A, B, C connected   A and B not connected

Electrical Switching Assignment 1

Using their knowledge of structures and gear systems, pupils produce a design, in their notebooks, for a vehicle which will run at about 1 m/min, given that the wheels are 35 mm diameter and the motor output-shaft speed is 250 rev/min. No construction is attempted at this time. Encourage pupils to draw sketches of a number of possible vehicle designs (at least three) and to select the best solution, giving reasons. The chosen solution should be drawn fully, including measurements. When constructed, the vehicle should be large enough to carry control equipment such as a relay unit and capacitor or a bistable unit. It is advisable, therefore, to suggest to pupils that the completed vehicle should be no smaller than, say, 25 cm long and 12 cm wide.

When all have completed the design stage, it may be desirable to have a general discussion. Also, make a typical calculation on the blackboard to emphasise that calculations should be recorded in a clear and logical manner. Before any construction can be attempted pupils must know:

a) the actual output-shaft speed, using a given gear ratio;
b) what Meccano gears are available.

It is possible, using the gear sections supplied with the motor, to obtain a gear low enough to drive the vehicle directly. But since at this stage of the course it is appropriate for pupils to be faced with the problem of designing and building a reduction system using Meccano gears it is suggested that only the 3:1, 4:1 and 5:1 ratios are fitted. At 12 volts this gives an output shaft speed of about 250 rev/min. This should result in groups producing different reduction systems depending on the size of their road wheels. Observation and discussion of the various solutions across the class will be beneficial.

Pupils are asked to think of a method of finding the motor speed. Most will no doubt use the motor a high ratio and count the number of revolutions in a given time, using a marker on the motor shaft. For this they will require a stop watch or stop clock. Other possibilities should be considered and, when all methods have been tried, the teacher should show that the shaft speed of a high-speed motor can be measured using a photocell and an electro-magnetic counter (see *Electrical switching follow-up 1* and Equipment Guide page 40). The maximum speed of operation of an electromagnetic counter is about 25 pulses per second, which corresponds to a motor speed of 1500 rev/min. For higher speeds the electro-magnetic counter can be replaced by an electronic counter. The teacher may wish to demonstrate some of the other possible methods.

The booklet 'Construction Guide number 2 "Tachometers" ', published by NCST, Trent Polytechnic, Burton Street, Nottingham has several appropriate circuits.

When pupils have collected together the parts for their vehicles, they may have wheels and gears of slightly different sizes from those used in their calculations. On the basis of their new information, (motor speed, wheel size, and gears) they should recalculate the theoretical vehicle speed. The vehicle should then be constructed and tested over a suitable distance.

The teacher should check each vehicle for rigidity and reliability, once completed, since it will be used throughout the Electrical-switching assignment programmes. Checks should be made especially on motor-mounting methods, to ensure correct meshing of gears under all reasonable load conditions.

Apparatus Required (15 pupils, 5 groups of 3)

assorted Meccano parts (these must include a sufficient number of wheels with rubber tyres to prevent slipping), spur gears, worm gears and axles

five stop watches or stop clocks

five 1 metre rules

chalk — for marking motor shafts and for marking out the start and finish of a test track.

Demonstrations

1   Select a gear ratio which makes it impossible to count the rotations of the output shaft. Show that such speeds can be determined using an electromagnetic counter and a photocell.

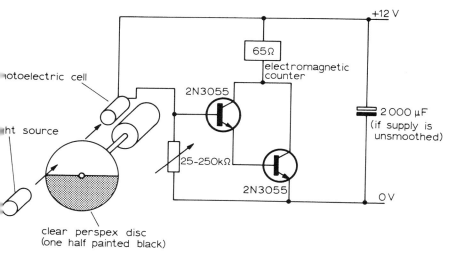

2　To count the rotations of a high-speed motor shaft, an electronic counter (e.g. a decade scaler) can be used.

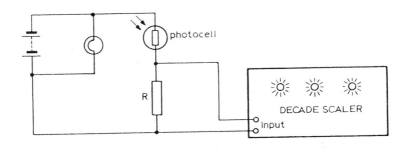

3　Demonstrate any other methods of determining motor-shaft speeds, e.g. using a stroboscope.

Buffer Experiments

1　Arrange for the vehicle to ascend an incline. Find the maximum angle of incline which the vehicle will travel up. Compare the result with other vehicles.

2　Arrange for the vehicle to ascend an incline. Add some extra weight and try again. What conclusions can be drawn? Repeat for other weights.

Homework Questions

1　A vehicle covers 2.5 m in 12 seconds. What is the average speed of the vehicle? How far will the vehicle travel in one minute at this speed?

2　A vehicle with 50 mm diameter wheels travels 4 m in 3 seconds. If the wheels are reduced to 40 mm diameter, what speed would you expect the vehicle to have?

3　When a vehicle travels over level ground, energy (electrical or chemical from petrol etc.) is consumed by the motor. What happens to this energy? If the vehicle climbs an incline, what happens to energy put into the motor?

4　Describe a method of finding the speed of rotation of a motor shaft which is too fast to measure by counting directly yourself.

5   Which of the following would *not* affect the speed of your
    vehicle:

    i)   the radius of the driving wheels?
    ii) the voltage supplied to the motor?
    iii) the gear ratio?
    iv) the axle diameter?
    v)   the diameter of the wheels which are not driven?

Electrical Switching Assignment 2

This is the first of several assignments concerned with the various
types of switches, e.g. manual switch, microswitch, reed-switch,
photocell, relay, etc. (Note that in *Electrical switching assignment 7*
the photocell is first introduced as a variable resistor but, in conjunction with a relay, is used by the pupils as a light-operated
switch.) Before attempting *Electrical switching assignment 2*, the
most we can assume is that pupils are aware of the single pole on-
off switch.

single-pole on-off

In this assignment we introduce switches with several poles and
several 'ways'. One could draw several types on the blackboard and
trust that the pupils develop a useful insight into switches in
general; however, this method will not have the same impact as
offering a new type of switch for them to investigate.

Pupils are given a two-way, two-pole (two-pole change-over)
manual switch unit, and are asked to discover which pairs of
sockets are 'joined together' with the switch in the 'up' and then in
the 'down' position. In effect we are calling on the pupils' knowledge of on-off switches and showing that there are four pairs of
sockets which become on-off switches. The 'short-circuit' detector
used is the vehicle motor in series with a 12 volt supply. This is far
more meaningful at this stage than offering the ohmmeter as a
means of detecting short-circuits. Obviously a lamp indicator unit
could be used.

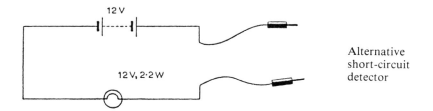

Alternative
short-circuit
detector

Having determined which pairs of sockets became short-circuited, pupils are asked to use the switch to isolate the motor completely from the supply, anticipating the following:

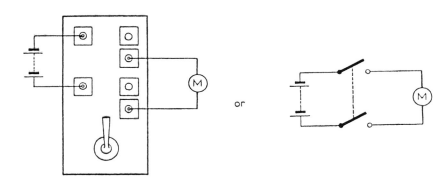

or

If the motor leads are in the position of the supply leads, and vice versa, the solution is just as acceptable.

At this stage the teacher should discuss two important points relating to mains switches.

a) By convention, a manual switch in the 'up' position should indicate an 'off' condition and in the 'down' position and 'on' condition (except on stairways). Unfortunate accidents can occur if this convention is reversed. (It is desirable, but not always possible, to switch off the entire mains supply in a building).

b) Again for safety, mains equipment should be completely isolated from the supply of using a two-pole on-off switch, though electrically a single-pole on-off is equally effective.

Reversing circuits are introduced by asking pupils to make their vehicle go in one direction with the switch in the up position and to reverse in the switch 'down' position. Although this appears to be a large step, experience has shown that the average pupils can solve the problem relatively quickly. This shows the wisdom of introducing the manual switch box in the manner suggested, thereby giving pupils complete insight into the two-pole, two-way switch. In order to assist pupils, but mainly to prevent excessive short-circuiting of the supply, pupils are instructed to place the motor leads into the yellow sockets (the switch poles). If difficulty is encountered, refer pupils to *Electrical switching assignment 2 (3)*.

Finally, pupils are asked to make use of a second manual switch unit so that they can make their vehicle stop — in addition to moving forwards and reversing. The obvious solution is to use the second unit to break a motor supply lead. It is interesting to note, however, that one or two groups frequently offer the solution below.

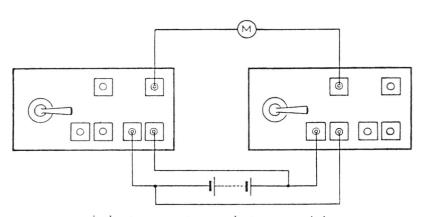

wired as two separate one-pole, two-way switches

In effect this solution prevents the two switches in a single switch unit from acting together, i.e. they have removed the 'ganging'. *Electrical switching follow-up 2* suggests that those who have not tried this method should do so. Later, in logic circuitry, pupils will meet this arrangement again in the form of a useful logic 'gate'.

Electrical switching follow-up 2 shows switches with differing poles and ways. These should be recorded for future reference. Note that shaded circles are used for 'pole' and open circles for 'ways'.

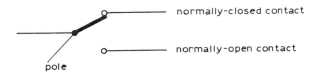

Apparatus Required (15 pupils, 5 groups of 3)

five vehicles fitted with a motor (made in *assignment 1*)

ten manual switch units (two-pole, two-way)

stackable 4 mm plug leads — assorted lengths

five power supplies — 12 V d.c. 2 A — with safety fuse or trip

Demonstrations

1   Discuss mains isolation using two-pole on-off switches

2   Show room-lighting switches as one-pole on-off. Discuss the danger of removing faulty lamps (e.g. broken glass) if neutral is switched rather than live.

3   Discuss 'earthing' of electrical appliances for safety — if the 'live' side of the mains shorts to an 'earthed' metal case, the fuse blows; if the metal case is not earthed, it becomes 'live'.

4   Establish the fact that one can place a switch in either the positive or negative lead from a battery, but that the mains 'live' lead should be switched.

5   Show how to wire up a mains plug.

6   Demonstrate several types of switches, including wafer switches, with assorted poles and ways.

Buffer Experiments

1   Wire up a reversing circuit which also provides for on-off switching using two separate one-pole, two-way switches, but

replace the motor with a lamp unit. In what way are the results using a lamp different from those obtained when a motor is used? Tabulate your motor and lamp results, and try to explain the differences.

2   Set up your circuit as in 1 and, using a lamp indicator, check that the lamp is on or off in the switch positions shown in your table. Now rewire the circuit such that the results are the reverse of those in your table.

Homework Questions

1   On a stairway there are usually two switches, one at the top and one at the bottom of the stairs. You will have noticed that, unlike any other switch in your house, you can sometimes put the stairway light 'on' by putting one of the switches up, but at other times the same switch must be down. Establish this on your stairway lighting at home, then show how you can do the same thing using a 12 volt supply, a 12 volt lamp unit, and two manual switch units.

2   Draw a diagram to show how you could use only two switches in a circuit with three lamps such that one, two, or three lamps can be switched on.

3   What kind of switch (i.e. one-pole on-off; two-pole, two-way; etc.) should be used to:

i)   operate a living-room light?
ii) operate a light on a stairway?
iii) switch an electric cooker on and off?
iv) switch a battery portable radio on and off?

4   Draw an electrical circuit diagram of this arrangement:

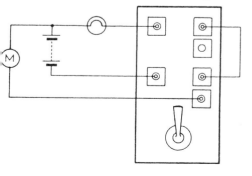

Describe what happens when
a) the switch is 'up',
b) the switch is 'down'.

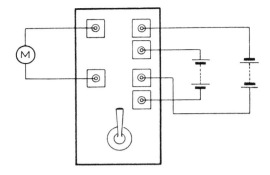

5   What is the purpose of this circuit? Draw an electrical circuit diagram of the arrangement.

Electrical Switching Assignment 3

This assignment introduces pupils to the microswitch and the reed-switch. A microswitch is required which is robust, will operate at low forces, and is of the single-pole, change-over type. The switch contacts are connected to three 4 mm sockets, to enable connections to be made quickly. The reed-switch should have heavy-current contacts and be of the single-pole, on-off, normally-open type. The reed-switch is fragile and must be protected against mechanical damage. The 4 mm sockets are connected to the switch contacts to facilitate quick assembly.

Should a heavy current be passed in the reed-switch contacts, brought about by shorting out the supply with the switch, welding of the contacts is likely to occur, preventing their opening when the magnetising force is removed. Usually the fault can be cleared by giving a sharp tap with a small screwdriver. If the contacts have been badly overheated they may be no longer lose all their magnetism when the actuating magnet is removed. This can result in a reed-switch that does not open correctly. Such a switch cannot be repaired. Again, large capacitors should not be discharged or charged rapidly through reed-switch contacts, since welding will occur. When pupils use a capacitor/relay delay in *Electrical switching assignment 5*, a microswitch with heavier contacts is used rather than a reed-switch. The microswitch contacts are able to arry the capacitor charging current, which can be very large at the nstant of switch closure.

Pupils are offered a microswitch in *Electrical switching assignment (1)* and are asked to test it to discover the type of switch used. It is nticipated that they will use similar tests to those used for the manual switch unit. The microswitch is attached to the vehicle, and no difficulty should be experienced in making the vehicle motor top when an obstruction operates the switch. When pupils draw his simple circuit diagram, it should be a conventional electrical circuit. To indicate a microswitch, a suitable method is as follows:

When it is convenient, introduce the use of the terms 'normally-open contact' — NO — and 'normally-closed contact' — NC. In ny switch which is biased one way (e.g. a microswitch or a relay), ircuit diagrams should be drawn showing the switch in the noperated position, as follows:

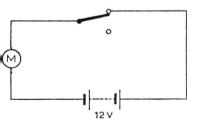

microswitch shown in unoperated condition, i.e. before vehicle strikes obstruction

n the case of a relay, the unoperated condition is when the coil does not pass a current.

Later, pupils examine a reed-switch. Each group should also be given two bar magnets. When the magnet approaches a reed-switch, a 'click' is quite audible as the contacts close. Before fitting the reed-switch to the vehicle, pupils should ascertain that it is a single-pole, on-off switch. Having easily arranged for a lamp to be lluminated, pupils will try to make the vehicle move with the reed-switch in series with the supply and will be unsuccessful unless a magnet is placed alongside the reed. It is important to know that

85

this on-off switch is 'normally-open', and the teacher should discuss this with each group as they are watched at work.

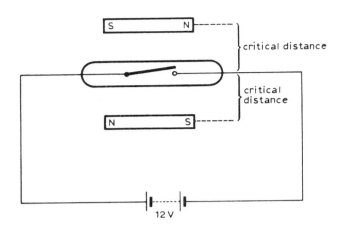

In order to make the vehicle stop when it approaches a magnet on an obstruction, the two magnets must be orientated such that they attract. The vehicle should stop at a short distance from the obstruction. All groups should be able to solve this problem after a hesitant start, but they will find that the reed biasing magnet position is very critical. It is intended that pupils should not be entirely satisfied with this arrangement, since the next assignment introduces a relay method which is far more reliable.

Apparatus Required

five microswitch units

five reed-switch units

ten strong bar magnets — poles marked (25–50 mm long)

five vehicles — constructed in *Electrical switching assignment 1*
stackable 4 mm plug leads — assorted lengths.

five power supplies — 12 V d.c., 2 A — with safety fuse or trip

five lamp indicator units.

five suitable obstructions — wooden blocks etc.

Demonstrations

1   Discuss the meaning of 'normally-open' and 'normally-closed' contacts. Show a dismantled microswitch to locate the biasing spring.

2   Show a selection of microswitches. Demonstrate that some require larger forces than others, and that the type of switch (poles and ways) can vary.

3   Show a selection of reed-switches. Discuss why smaller types have smaller contacts and cannot pass large currents. If possible, demonstrate a single-pole change-over type. Set up suitable circuits to demonstrate these.

4   Demonstrate the use of a coil to provide a magnetic field for the reeds.

e.g.

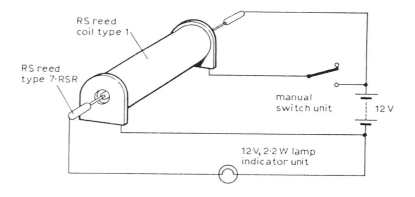

Fit a reed-switch to a device such as a lathe bed. Fit a bar magnet to the carriage. Arrange for a lamp to illuminate when the magnet approaches the reed. Discuss whether the reed used as a 'limit switch' would be suitable to give a warning to the operator.

uffer Experiments

'ire up your vehicle motor to a manual switch unit so that you can ᴑake the vehicle move forward or reverse. Fit a microswitch to ᴑch end of your vehicle. Somewhere on the vehicle fit a lamp ᴑdicator unit. Arrange for the lamp to be illuminated when the vehicle strikes an obstruction and, each time it illuminates, reverse the motor with the switch box. Draw your circuit before actual wiring up.

Homework Questions

1 Manual push-button on-off switches are available from certain manufacturers. The button is maintained in the up position by spring, and the switch is operated by pushing against it. These switches are 'normally-open' or 'normally closed' Imagine you manufacture these two types of switch and write a brief explanation of each.

2 Describe what happens in the circuit below if one or both of the magnets A and B are placed next to the change-over reed-switches (A and B). Assume that in the diagram the magnets are too far away from the reed-switches to operate the contacts.

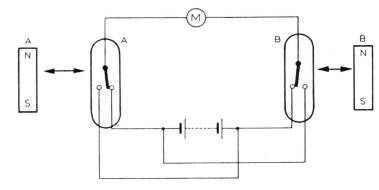

3 The intention with the circuit below is for the lamp L to be switched off when a magnet is placed near to the reed-switch, provided that switch S is closed.

   i) Is the lamp switched off?

   ii) Why is the circuit not a good one?

   iii) Redraw the circuit to show your modifications.

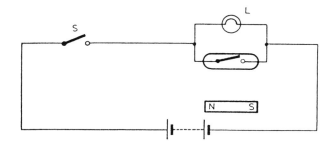

4   The diagram shows the type of microswitch which is intended to
    operate when a force F is applied at the end of the rigid bar as
    shown. Unfortunately, F is too small to operate the switch and
    cannot be made larger.

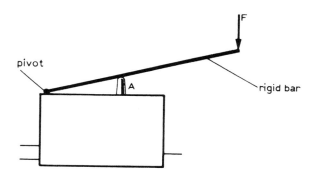

i)   What could you do to enable the force F to operate the
     switch reliably?
ii) The force F required to just operate the switch is 0.1 N. The
     distance of the pivot from A is 20 mm and the distance
     from the pivot to F is 100 mm. What force applied directly
     on the plunger A would just operate the microswitch?

5   You wish to use a microswitch method to switch on a d.c.
    electric motor requiring a starting current of 3 A. Unfor-
    tunately, although you have several identical microswitches, the
    contact current rating of each is 1 A d.c. maximum. How
    would you solve the problem?

Electrical Switching Assignment 4

One of the aims of the *Control Technology* course is to encourage
inventiveness. This is possible only if there are problems to solve
and the materials with which to solve them. *Electrical switching
assignment 5* brings the pupils' first real acquaintance with the electromagnetic relay. This important control device is extremely versatile and can be used in a wide range of applications. The pupils
meet several useful relay circuits in the basic course, and teachers
will no doubt find that the more able ones will suggest other con-
figurations to solve individual problems.

The manual switch unit was introduced by testing to find out which
pairs of sockets became shorted out when the switch was operated.
The relay is examined and pupils soon recognise that the two units
are similar. Pupils are instructed to connect a 12 volt supply across
the green sockets (relay-coil terminals) and to note the effect by
observing the relay operation. This is to establish that the relay
'changes over'. If pupils have not met relays previously, then this is
the time to demonstrate several types, and to discuss the principle
of operation. Having completed their initial tests, pupils should be
left in no doubt that they are now dealing with a two-pole, two-
way, coil-operated switch.

The relay is used to control the vehicle and, in the first instance, is
employed to *invert* the switch function, in the reed-switch unit.
Being normally-open, a reed-switch will not, unless biased with a
magnet, allow a vehicle motor to run. By putting the reed-switch in
series with the relay coil, the relay can act as an 'inverter' or
'buffer' as shown.

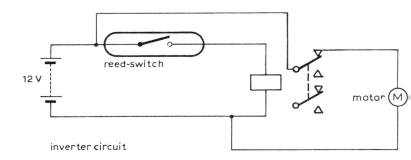

90

he reed-switch is normally-open and remains open until it is in
lose proximity to a magnet. The motor is connected to a pair of
ormally-closed contacts in series with the supply, and therefore
uns. When the reed-switch closes, due to the proximity of the
nagnet, the relay energises and disconnects the supply to the
notor.

Jote that the only relay contacts required are one-pole on-off, the
econd set of contacts being left disconnected in this case.

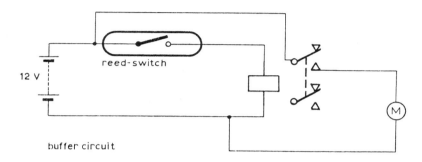

the relay contacts used are normally-open, as shown, the circuit
electrically analogous to a reed-switch alone, and a magnet is
quired to start the motor.

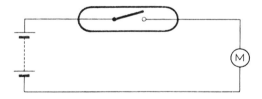

n advantage of this relay circuit is that the motor current is
rried by the relay contacts and not by the reed contacts. The
tter, therefore, carry only the small relay-coil current. The
aximum current permissible with most reed contacts is quite small
ess than 1A in general), whereas relay contacts can switch several
nperes. Experience has shown that for reliable intermittent use the
aximum d.c. current is in the region of 0.5 amperes for the reed-
vitch suggested. This is the approximate starting current for a
asonably loaded motor. It is therefore permissible to switch the

motor current by means of a reed-switch unit. If a larger motor were to be used, or a reed-switch with a lower contact-current rating, then a relay buffer would be essential.

Pupils should be made aware of these two uses of a relay at the earliest opportunity, and, if the teacher thinks it appropriate at thi stage, he should now discuss the two applications. Note that the terms 'buffer' and 'inverter' are not mentioned in the Electrical-switching assignments, but pupils should be made familiar with them.

The final investigation of Electrical-switching Assignment 4 is concerned with finding a way of making the vehicle reverse away from an obstruction fitted with a magnet. Pupils should find little difficulty in wiring up a reversing circuit to the relay contacts, and should succeed in making the vehicle reverse. Unfortunately, as soon as the reed-switch is no longer influenced by the magnet on the obstruction, the vehicle again reverses towards it. The sequence repeats itself and the vehicle 'oscillates'.

Apparatus Required (15 pupils, 5 groups of 3)

five reed switch units

five two-pole change-over relay units

stackable 4 mm plug leads — assorted lengths

five vehicles fitted with motors — made in *Electrical switching assignment 1*

five power supplies — 12 V d.c., 2 A — with safety fuse or trip

Demonstrations

1   Demonstrate the relay functioning as a 'buffer' and as an 'inverter'.

2   Demonstrate the operation of various relays, showing differen sizes of coils and different types of contacts. Some British Telecom types are particularly suitable for discussing the principle of operation.

3   Demonstrate and discuss the uniselector principle and applications (see following notes).

4   Demonstrate the operation of an electromagnetic counter, and discuss its similarity to a relay.

Buffer Experiments

1 Connect up the relay unit in order to make the vehicle motor reverse when the relay is energised. Replace the reed-switch unit by a microswitch unit. Compare the results obtained when the microswitch strikes an obstruction with those using a reed-switch. Try to account for any differences in vehicle operation.

2 It is possible to make a vehicle reverse away from an obstruction, and continue reversing, by wiring up the relay in such a way that once it energises it remains so until the supply is switched off. Attempt to set up such a circuit, and draw it in your notebook.

Homework Questions

1 Draw an electrical circuit diagram for the arrangement shown opposite, and describe what happens when switch A is opened and closed.

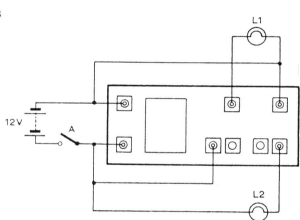

2 Describe what is likely to happen in the circuit below, when the switch S is opened and closed.

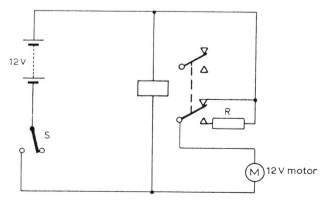

93

3   Draw an electrical circuit diagram of a device which will enable
    a motor to run forwards, and reverse when an on-off switch is
    operated. When the motor runs forwards, a lamp (call it L_1)
    should illuminate; and when it reverses, a second lamp (L_2)
    should illuminate and L_1 should go out.

4   A motor, which operates from the mains 240 V a.c. supply, is
    situated about 500 m from a building. It must be possible to
    switch the motor on and off from the building. Permission is
    refused to allow mains high-voltage cables to run from the
    building to the motor, owing to the possible danger to the
    public. If the motor is situated close to a mains supply, show
    how you would solve the problem.

5   Draw a diagram of an electromagnetic relay with one-pole, two-
    way (single-pole, change-over) contacts. Label the important
    parts and describe how the relay operates.

The Uniselector

The uniselector is a multi-position relay which is sometimes referred
to as a 'stepping' relay. The common two-position relay is a
versatile device for pupils, but the uniselector broadens the field of
applications much further and is, therefore, worth including in the
course. No assignments dependent on this device have been
prepared, owing to its close relationship to the relay. It is desirable,
however, that pupils should see one or two different types to
compare physical size and number of positions. Experience has
shown that, if uniselectors are available, pupils will make use of
them in solving problems arising during project work.

This shows a circuit diagram of a two-bank, ten-way (two-pole, ten-
way) uniselector. If a supply is connected across the coil AB, the
armature is actuated in a similar manner to that in a relay. This
action causes a pawl to move to the next tooth on a ratchet wheel.
When the supply is removed, a spring, which has been extended
whilst the coil was energised, has the effect of pulling the switch
over to the next position, since the switch wipers are attached to the
same shaft as the ratchet wheel.

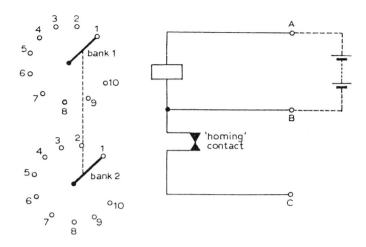

The spring steel latch prevents ratchet rotation when the coil energises but allows rotation under the action of the spring as the coil de-energises. A length of phosphor-bronze wire, or similar, rubbing on the wiper shaft serves as the pole of the switch.

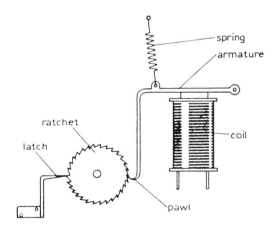

The switch is continuous in action in that, in this case, when wiper X leaves contact position 10, wiper Y connects with contact position 1. If the uniselector has two banks, a second pair of wipers X_1 and Y_1 are fitted to the same shaft as wipers X and Y but are insulated from them. Uniselectors having either eight banks and twenty-five ways or four banks and fifty ways are quite common.

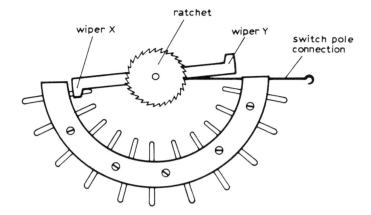

Since the switch rotates continuously in only one direction, some means of quickly returning the uniselector wipers to contact position 1 is obviously desirable. This is achieved by making use of the 'homing' contacts. These contacts are operated by the armature.

When the coil is de-energised, the 'homing' contacts are closed; but when the coil is energised, mechanical linkage from the armature opens them. Thus, if the coil is supplied through these contacts, the armature will oscillate, as the contacts make and break, and the switch will rotate at a rapid rate.

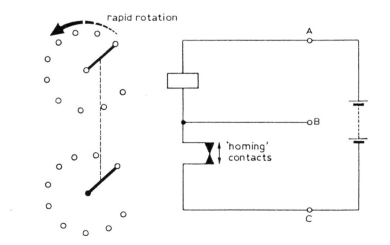

In order to stop the switch at position 1, one bank of the uniselector is used as a 'homing' bank. All contacts except contact position 1 are connected together and wired as shown.

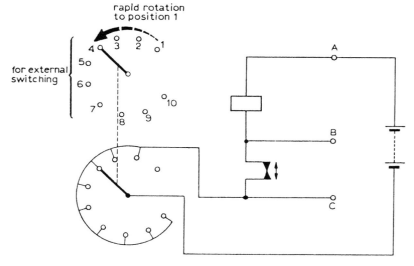

Using the same technique, the switch can be stopped at any position, or even at a number of positions. A typical use of 'homing' is shown below.

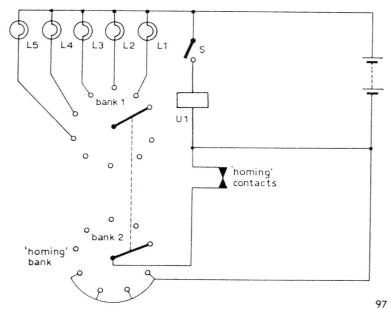

With the uniselector in position 1(home) as shown, all lamps L_1 to L_4 are extinguished. When switch S is impulsed, the uniselector moves to position 2, illuminating L_1. At the next impulse, L_1 is extinguished and L_2 illuminates; and so on to L_5. At the sixth impulse the device 'homes' to position 1.

A uniselector unit

A uniselector enclosed in a Perspex box with the pole contacts and coil connections wired to 4 mm sockets is a useful addition to the basic equipment. A small uniselector of three banks and ten–twelve ways is particularly suitable. If the uniselector has more positions than this, internal 'homing' beyond position 10 could be arranged.

A typical layout:

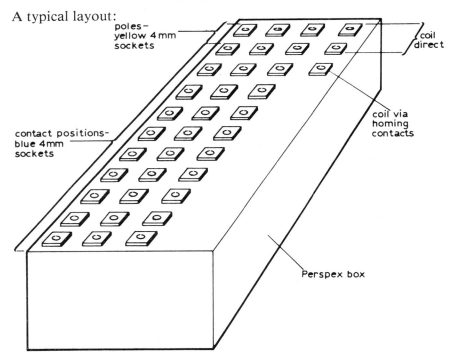

A homing bank can be conveniently arranged externally by linking switch positions by short leads fitted with stackable 4 mm plugs.

Electrical Switching Assignment 5

Capacitors can be used effectively for 'delaying' the operation of relays, and this assignment sets out to introduce pupils to this particular function of a capacitor. In the assignments, the capacitor is dealt with not from the standpoint of theory and construction, but in terms of its usefulness in control mechanisms.

Pupils observe what happens when a relay is connected across a supply and then disconnected. This shows quite clearly that the relay responds relatively quickly to the 'making' and 'breaking' of the supply. A capacitor is now fitted across the relay coil. When the supply is connected, the relay again energises in a very short time.

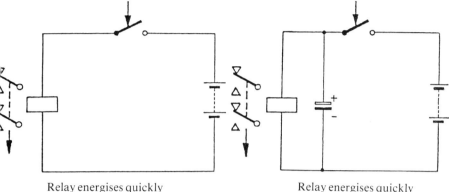

Relay energises quickly   Relay energises quickly

There is no discernible difference in the time taken for the relay to energise. When the supply is disconnected, however, the relay with the capacitor wired in parallel remains energised for a considerable time.

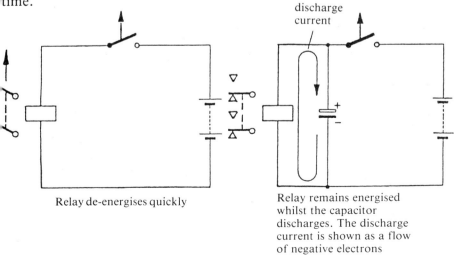

Relay de-energises quickly

Relay remains energised whilst the capacitor discharges. The discharge current is shown as a flow of negative electrons

99

Pupils will quickly realise that the capacitor stores electricity (electrical charge) when the switch is closed and gives up the charge to the relay coil over a short period of time after the switch is opened. At the instant the switch is opened the capacitor is already charged and has 12 volts across its terminals. As it discharges the voltage falls. Eventually the voltage across the capacitor plates is too low to drive sufficient current through the relay coil to keep it energised. An alternative explanation is that, at the beginning of the discharge, the discharge current is high, and this falls off with time. At a certain low current the relay de-energises. Theory and experiment show that the discharge current falls off exponentially with time.

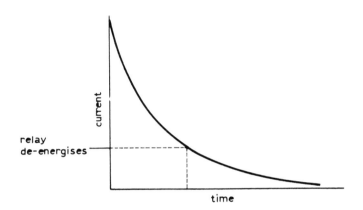

We do not mention this exponential decay yet. However, students from now onwards will be using capacitors regularly, and at some time the teacher should demonstrate the simple experiments mentioned in *Appendix 3 — Capacitors.*

Having established that a capacitor can 'delay' the de-energising of a relay, pupils investigate the effect of using different values of capacitor on delay time. Using the model vehicle, pupils measure the distance that their vehicle reverses using, say, 2000 μ, 5000 μF, and 10 000 μF capacitors. The reversing distances should be approximately proportional to the capacitance. The values are not critical provided that they are whole-number multiples of the lowest value. Capacitors of less than 500 μF are likely to give only very short reversing distances.

The reversing distances obtained when capacitors are connected in series and in parallel will indicate to the pupils that the effective value of two capacitors connected in parallel is the sum of the two values. If two capacitors of equal value are connected in series, the effective value is one half of the value of one of them. Pupils may try connecting different value capacitors in series, but we should not expect them to deduce

$$\frac{1}{C} = \frac{1}{C_1} + \frac{1}{C_2} + \text{--------} \frac{1}{C_n}$$

Points to Note

1   Pupils are warned in *assignment 5* to ensure that the capacitors are connected the correct way round, i.e. the positive terminal of the capacitor to the supply positive and the negative terminal to the supply negative.

2   The values of capacitors should be clearly marked. It is not necessary to define the 'farad', but pupils will discover from their experiments that, the greater the number of microfarads (μF), the larger the charge.

3   Pupils should not use reed-switches in circuits which include a capacitor across a relay coil. When a discharged capacitor is connected across a supply, the initial charging current is limited only by the resistance in the circuit. If this resistance is very low, the initial (surge) current can be in the order of several amperes. In the circuit shown, the critical current is high (since the total resistance — external — wiring resistance plus supply internal resistance — is very low) and the reed contacts are likely to become welded together.

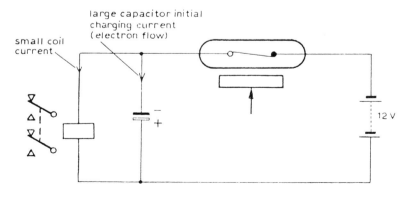

The same argument is applicable to capacitor discharge, large currents flowing if the external resistance is low.

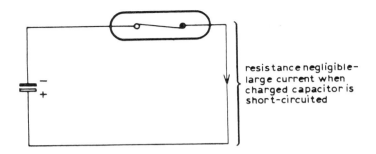

resistance negligible-
large current when
charged capacitor is
short-circuited

Because of the large charging currents, a microswitch with heavier contacts is used in preference to a reed-switch in *Electrical switching assignment 5*.

4   The energising current for a relay is much higher than its 'holding' current.

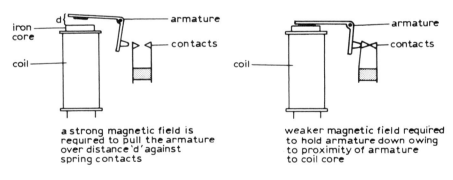

a strong magnetic field is
required to pull the armature
over distance 'd' against
spring contacts

weaker magnetic field required
to hold armature down owing
to proximity of armature
to coil core

This holding current varies not only between different types of relay, but also between apparently identical types: the holding currents of the relays suggested may vary by several milli-amperes. The consequence of this inconsistency is that, for the same value of capacitor, pupils will find other vehicles reversing for different distances. (Variations may also be due to frictional factors in the model vehicles.)

5   The delay time of a relay is a function of its coil resistance in addition to the capacitor value employed; thus longer delay times can be achieved using relays with high resistance coils.

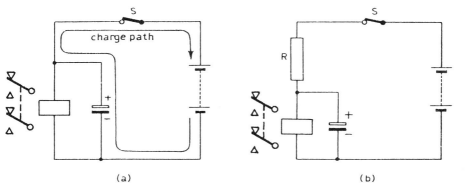

(a)   (b)

Electrical switching assignment 5 introduces a 'delay' caused by
a relay remaining *energised* after the supply is removed. Relays
can be made to remain *de-energised* for a short period after the
supply is connected. The method is much more critical, but it
may be of use in project work. A series resistor is employed.

When there is no additional resistance *R*, see diagram (a), the
capacitor charges rapidly when the switch S is closed, owing to the
low circuit resistance. The uncharged capacitor has very low
resistance initially, and 'short-circuits' the relay coil, but for only a
very short time. In diagram (b) the capacitor now takes longer to
charge, because of the additional resistance *R*. When the voltage
across the capacitor plates has risen, the requisite energising current
passes through the relay coil, and the relay energises.

By making the resistance *R* variable, the delay can be increased and
decreased. If *R* is made too large, however, its resistance limits the
circuit current to less than the relay energising current, and the
circuit ceases to function.

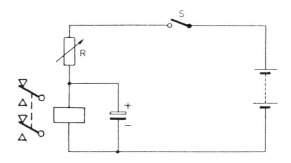

Calculation of series resistance R

The recommended relay has a minimum energising current of approximately 20 mA and a coil resistance of 185Ω

Total circuit resistance = $R + 185Ω$

If the supply voltage is 12 V and the current required is 20 mA,

$$R = \frac{V}{I}$$

$$\therefore R + 185 = \frac{12}{20/1000}$$

$$\therefore R = \frac{12\ 000}{20} - 185Ω$$

$$\therefore R = 600 - 185Ω$$

$$= 415Ω$$

Thus the maximum value for R is 400 Ω (approx)

If this value is exceeded the relay cannot energise. A 500 Ω variable resistor could be used to adjust the delay time.

Apparatus Required (15 pupils, 5 groups of 3)

five reed-switch units

five microswitch units

five relay units (two-pole change-over)

five 2000 μF capacitor units

five 5000 μF capacitor units

five 10 000μF capacitor units

several miscellaneous capacitor units

five vehicles fitted with motor — made in *assignment 1*

five power supplies — 12 V d.c., 2 A — with safety fuse or trip

five 1 metre rules or similar

stackable 4 mm plug leads — assorted sizes

Demonstrations

1   At an appropriate time, the charge and discharge characteristics of capacitors should be studied in more detail — see *Appendix 3 — Capacitors.*

2   Discuss the effect of connecting capacitors in series and parallel in the light of *Electrical switching assignment 5.*

3   Discuss factors affecting the energising current and minimum holding current for electromagnetic relays.

Buffer Experiment

Measure the vehicle reversing distance when capacitors of different values are used. Plot a graph of capacitor value against distance.

Homework Questions

1   Draw a circuit diagram to show how you could arrange a lamp to be illuminated when a spring-loaded push switch is operated and to remain illuminated for several seconds after the switch is released.

2   In each case in the circuits shown below, the switch S is operated for a sufficient time to allow the capacitor to fully charge.
    Which capacitor is storing the greatest charge and which the least?
    What is the ratio of the charges stored by C_1, C_2, and C_3?

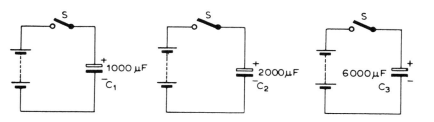

3   Indicate how the current flows in the circuit shown below at the instant the switch S is closed. Indicate the current flow in the circuit after the switch has been closed for some considerable time.
    Give reasons for any answers you give.

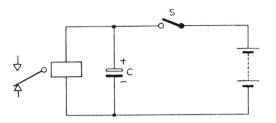

4   In the circuits below, RL1 and RL2 are identical relays. Which relay remains energised for the longer time after switch S is opened? Give a reason for your answer. (If you are not sure of the answer, set up the equipment and test it.)

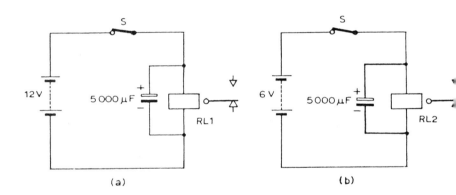

(a)   (b)

5   What is the effective capacitance in each of the following arrangements?

(a)

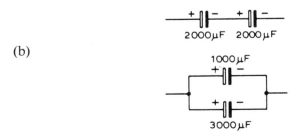

(b)

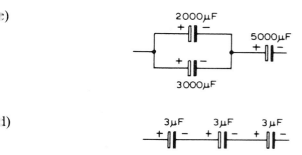

:)

2000μF

5000μF

3000μF

1)

3μF   3μF   3μF

lectrical Switching Assignment 6

ater in the course, pupils will work on assignments involving logic,
•gic circuits, and some important items of ancilliary switching
quipment known as *astable, monostable,* and *bistable* devices.

√hilst these devices are often used in association with other logic
rcuitry, they are useful items for use separately in the control of
quipment and operations. They are commonly referred to in this
ay:

astable — 'clock'

monostable — 'delay'

bistable — 'flip-flop'

lectrical switching assignments 5 and *6*, with suitable demonstra-
ons, deal with these three functions. Each device can be identified
y its number of stable 'states', i.e. none, one, or two. For
.ample, a manual switch can be up or down, a relay energised or
ot energised, a transistor may be conducting or not conducting.
'e can simply call the states 'on' or 'off'.

t some switching systems, however, neither state (on or off) is
able, and the system automatically switches itself from one state
• the other continuously. Such a system is referred to as *astable*
without stability). The to-and-fro excursions of an astable device
eat out time at a definite frequency; the device is therefore
halogous to the ticking of a clock. Pupils have already met a kind
f astable system in the form of the oscillating vehicle. By
ompleting *assignment 6*, pupils are introduced to a relay oscillator
efer to the notes on 'a double-relay oscillator' later in this
ction).

107

A *monostable* system has one
stable state, as the name
suggests. Again, pupils have
met such a system when a
capacitor is fitted in parallel
with a relay coil.

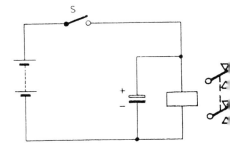

If an 'impulse' is applied by operating a spring-loaded switch S, the
relay energises and the capacitor simultaneously charges. When S is
opened, the relay remains energised for a time dependent upon the
charge stored in the capacitor. The stable state in this case is when
the relay is not energised, remaining so until an external impulse is
applied. The energised relay is then in an unstable state, since it
automatically de-energises when the capacitor has discharged
sufficiently. The term 'delay' is applicable to this system, since
there is a delay between the releasing of switch S and the
de-energising of the relay.

A *bistable* device has two stable states. Perhaps the simplest
bistable mechanism is a manual on-off switch such as a room
lighting switch; the switch remains indefinitely in one or other state.
The term 'flip-flop' is sometimes used because, when an impulse is
applied, the device 'flips' into one stable state but 'flops' into a
second stable state at the next impulse.

Electrical switching assignment 6 introduces pupils to a useful
bistable relay device.

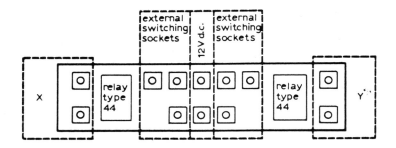

A double-relay circuit is employed for three main reasons:

a) the inputs are in effect short circuits applied across X or Y — the possibility of damage to the unit is therefore less than if a supply voltage were used;

b) both inputs are identical, thus any method used to give input X can also be used to give input Y;

c) the circuit is closely analogous to transistor bistable circuits.

Method of operation

bistable unit — circuit diagram 1

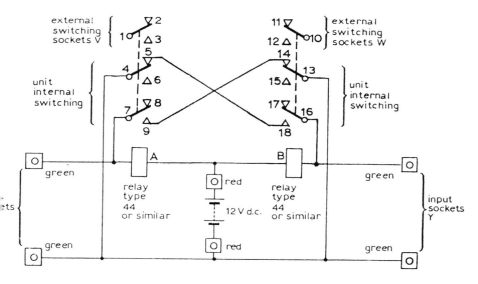

NOTE: the numbers used are for the purpose of this description only and do not refer to any numbers printed on the relay type 44, nor to the numbered sockets, as given in the pupils' notes.

When the supply is first connected, both relays are de-energised (see circuit diagram 1). A short-circuit applied at X will cause relay A to energise, while a short-circuit applied at Y causes relay B to energise. Assuming sockets X are short-circuited, relay A energises and remains energised (even with the short-circuit removed) since its own contacts 7 and 9 are now joined, connecting the left-hand side of the delay coil A to supply negative through relay B closed contacts 13 and 14. The right-hand side of relay coil A and the left-hand side of relay coil B are permanently wired together and connected to supply positive (see circuit diagram 2).

Any further impulses applied at X have no effect. Note that the external switch sockets (V) have now changed connections. External switch sockets (W) remain as before.

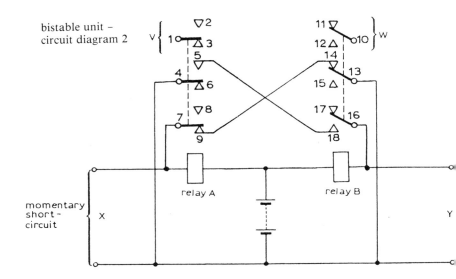

2   If a short-circuit is applied at Y, relay B energises. Connection between contacts 13 and 14 is broken, causing relay A to de-energise, having lost its connection to supply negative. When relay B energises, 16 is connected to 18; and, when relay A de-energises, 4 is connected to 5, thus connecting relay coil B to supply negative. If the short-circuit at Y is now removed, relay B will remain energised (see circuit diagram 3).

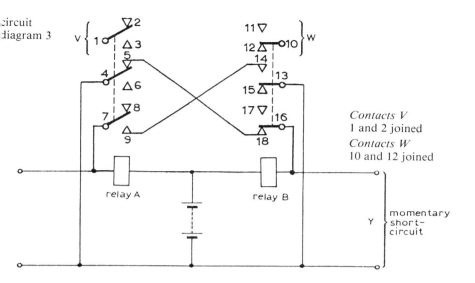

Contacts V
1 and 2 joined
Contacts W
10 and 12 joined

A further short-circuit applied to Y has no effect. However, a short circuit applied at X causes both relays to change over again to the situation shown in circuit diagram 2, and so on.

Notes.

1 Either relay when on (energised) or off (not energised) is in a stable state.

2 The supply polarity can be reversed without affecting operation, i.e. the supply sockets need not be indicated positive and negative.

3 Except when the supply is first connected (or when there is no supply connected), one relay, and one only, should be energised (see note 4).

4 Inputs must be momentary short-circuits. If, for example, X remains short-circuited and a short-circuit is applied at Y, then both relays energise. This situation should be avoided, since the bistable action is lost.

5 An illuminated photocell, microswitch, etc. will operate as short circuits for the unit. In the case of a photocell, the 'impulse' must be caused by light striking the cell, *not* the action of breaking the beam to the cell.

6   The terms 'set' and 'reset' are commonly used when referring to bistables. When a short-circuit is applied at X, the bistable can be said to be 'set'; and when a short-circuit is applied at Y, it can be said to be 'reset'; and *vice versa*. These terms have little meaning unless the device is connected into an external circuit. The terms would have value, no doubt, in this case:

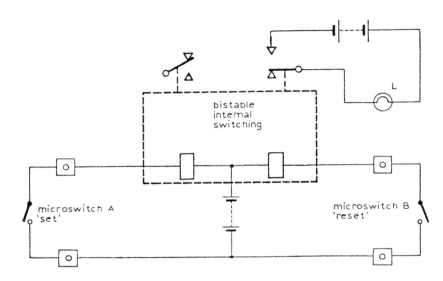

If microswitch A is operated momentarily, both relays change over and the lamp L is illuminated. Switch A has 'set' the circuit. To 'reset' the arrangement (i.e. to extinguish the lamp), the reset microswitch B is impulsed.

Content of assignment 6

Pupils investigate the operation of the bistable unit in much the same way as they investigated the manual switch unit and relay unit, and summarise their findings in tabular form. Once familiar with the switching action, the pupils should be able to use the external switch sockets to make a vehicle reverse. The majority will no doubt at first fail to take into account that the situation is not quite the same as for relay switching. In a two-pole, two-way relay, *both* switches operate together (ganged); this gives the reversing circuit shown.

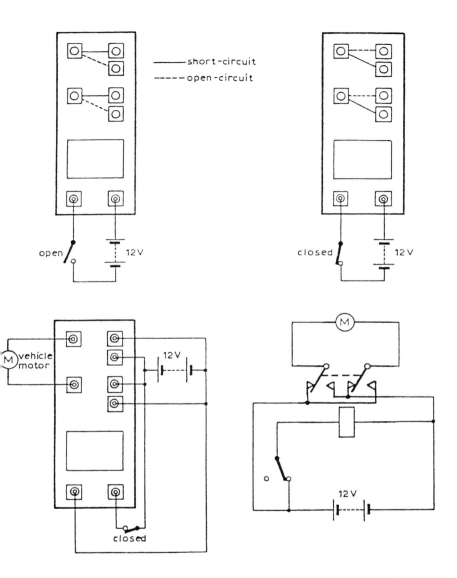

In the case of the bistable unit, either relay A or relay B is energised, which results in the external switching sockets being used as shown.

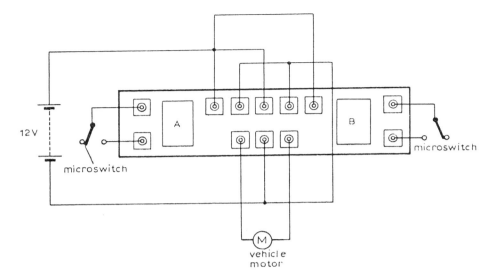

Most pupils should eventually appreciate the different arrangement required after one or two unsuccessful attempts. If necessary, a hint or two could be provided, but the solution should *not* be given. Pupils should be asked to state what must be done to achieve reversal of a d.c. motor, and they should be able to solve the problem once they are clear as to what is required.

Alternative assymmetrical bistable (latched relay)

A bistable using one relay is possible, but the impulse inputs 'set' and 'reset' are not identical.

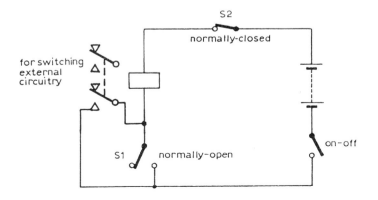

When the battery switch is operated, S_1 is closed momentarily, and the relay energises. One set of its contacts immediately places a short-circuit across S_1, allowing the latter to be opened while leaving the relay energised. This could be the 'set' condition. To reset, the normally-closed switch S_2 is opened momentarily. This de-energises the relay and removes the short-circuit across S_1. When S_2 again closes, the relay remains de-energised until S_1 is again operated.

A two-relay oscillator

Although not included in the assignments, teachers should, if at all possible, allow pupils to attempt to make a relay oscillator. The problem could be stated as, 'Design and construct a circuit in which two relays form an oscillator by making and breaking the supply to each other.' Capable pupils will probably succeed in solving the problem. Each relay can operate a lamp to indicate time intervals when energised (or de-energised).

Pupils who do not attempt the design of an oscillator should at least see that the arrangement is possible (see *Demonstrations*).

Circuit for a two-relay oscillator:

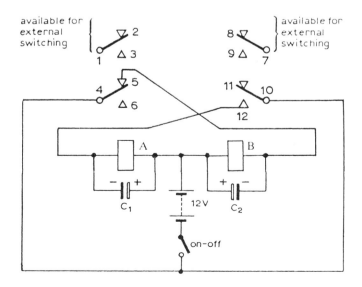

Relays A and B are two-pole, two-way relay units. The values of capacitors C_1 and C_2 determine the switching frequency (frequency of oscillation), and values from 100 μF to several thousand microfarads can be tried. The effect of using unequal capacitor values for C_1 and C_2 should be noted. Note that a pair of normally-closed contacts are used for one relay and a pair of normally-open for the other.

Method of Operation

When the on-off switch is closed, relay B energises immediately (and C_2 charges), since the right-hand side of relay coil B receives supply negative through contacts 4 and 5. Contacts 10 and 12 of relay B now close, thus energising relay A immediately (C_1 charges). With relay A energised, contacts 4 and 6 close, cutting supply negative to relay B. Relay B remains energised, however until C_2 has discharged. Relay B then de-energises, disconnecting the supply to relay A as contacts 10 and 11 close. Relay A de-energises only when C_1 has discharged. When relay A de-energises, relay B energises immediately through contacts 4 and 5, and so on. Note that the supply polarity should not be reversed, otherwise the capacitors are wrongly connected.

Single-relay oscillators

Single-relay oscillators are possible. A teacher may wish to investigate one or two and discuss them with pupils.

1   The speed of oscillation is limited by the mechanical construction of the moving parts (inertia).

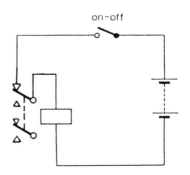

2   Large 'mark-space' ratio,
    i.e. the relay is on for a time
    determined by capacitor C
    but off for a very short
    time.

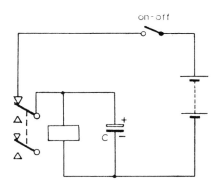

3   Capacitor C keeps the relay
    energised for a period of
    time as in 2. The resistor R,
    however, delays the charging
    of capacitor C, thus
    delaying the energising of
    the relay. Equal mark-space
    ratios are possible, but, if
    the value of R is too high, it
    will prevent sufficient
    current flowing in the coil to
    energise the relay. (Values
    for approximately one-
    second oscillation; R, 500Ω
    variable; C, 5000 μF; relay
    type 40 in two-pole, two-
    way relay unit.)

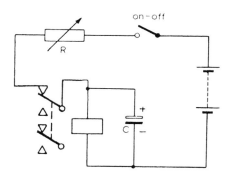

What do we expect from pupil involvement in assignment 6?

This will largely depend on the ability of pupils, but the following is
offered as a guide:

a) They should be aware of the bistable principle and understand
   the internal switching of the bistable unit as shown in the
   assignment 6 follow-up. They should not necessarily be expected
   to work out the circuit for themselves at this stage. They should
   be discouraged from attempting to *memorise* this or any other
   circuit; one aim of the course is to encourage pupils to design
   circuits by applying principles.

b) The pupils should make sufficient notes and diagrams to enable them to refer back to the principles involved should they require a bistable circuit as a control device in the future.

c) The terms 'set' and 'reset' should be introduced here and the associated concepts consolidated in later work.

d) Pupils should be aware of the single-relay bistable as a simple and cheap device which 'makes and breaks its own supply'.

e) They should be encouraged to draw a simplified diagram to represent a bistable unit when the latter is used as part of more complex circuitry. At first, pupils may wish to use a simple diagram representing the unit, for example:

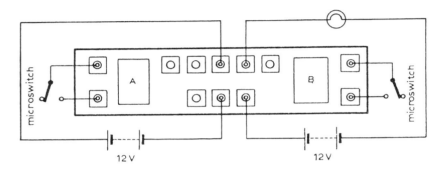

However, they should be encouraged to use a more general circuit such as:

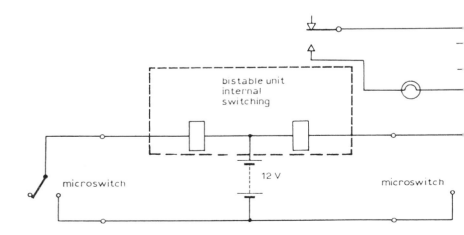

They should not be expected to draw the complete bistable circuit at all times, although a teacher may ask for this on infrequent occasions to prevent the unit becoming a 'black box'.

) Pupils should be able to recognise and construct a double-relay oscillator circuit and be able to explain how it works.

Apparatus Required (15 pupils, 5 groups of 3)

ive relay bistable units

en microswitch units (reed-switch units may also be used)

ive vehicles fitted with motor, as made in *Electrical switching assignment 1*

ive power supplies — 12 V d.c., 2 A — with safety fuse or trip

tackable 4 mm plug leads — assorted lengths

en suitable heavy obstructions, e.g. large blocks of wood, heavy veights, etc.

Demonstrations

Discuss the terms 'astable', 'monostable', and 'bistable', with the help of suitable equipment.

2 Discuss 'set' and 'reset' by assembling a suitable circuit, including two push-button switches (microswitch units held in the hand are suitable).

8 Demonstrate a single-relay bistable using a two-pole, two-way relay unit. (This could be investigated by pupils in order to produce, say, a burglar alarm.)

4 Discuss ways of drawing a bistable unit which is included in other circuitry.

5 Discuss the effect of applying a short-circuit to both pairs of green sockets. Show that the bistable action is then lost.

6 If pupils have not already attempted to task of producing a relay oscillator, the teacher should demonstrate and provide a circuit. Each group should then wire-up a similar unit.

Buffer Experiments

1 Attempt to design and wire-up a bistable unit using a single relay. Use a two-pole, two-way relay unit and two microswitches. The relay should energise when one microswitch is pressed, and should remain energised when the microswitch is released. The second microswitch should de-energise the relay when pressed, and the relay should remain de-energised when the microswitch is released.

Use a lamp indicator unit to assist you in determining whether or not the relay is energised.

Homework Questions

1 By reference to a circuit diagram (you will have one in your notebook, or use the circuit in *Electrical switching follow-up 6*), describe how the two-relay bistable works.

2 What are meant by the terms 'astable', 'monostable', and 'bistable' in electrical and electronic circuitry.

3 What happens in the circuits (a) and (b) if the switches S are closed? Give reasons for your answers.

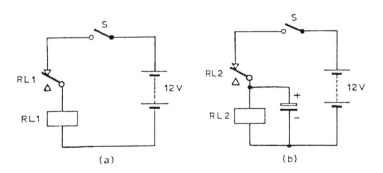

(a)   (b)

4 Draw a circuit diagram of a relay bistable device, using only one relay, and explain how it works.

Electrical Switching Assignment 7

The photocell used in these assignments is of the light-dependent resistor type, the Mullard ORP12 being ideal. Although in later assignments the photocell is used almost exclusively in a switching capacity, this assignment establishes that, as the light falling on the cell drops in intensity, the resistance increases, and *vice versa*. This change in light intensity is achieved by varying the distance between the light source and the photocell and noting the effect of a lamp indicator unit placed in series with the cell. If desired, a second indicator unit could be fitted directly across the supply for comparison purposes.

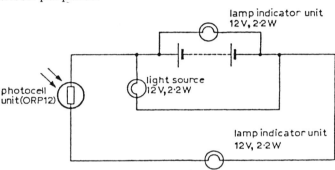

a)

Experience has shown that a 12 V, 2.2 W indicator lamp in series with the photocell does not give very satisfactory results. Much more impressive results are obtained if you supply the pupils with 5 V, 0.36 W LED indicator bulbs. These will not fit into the 'normal' bulb holder.

It is recommended that you obtain a set of these bulbs and holders. Fit the holders with two short lengths of stiff single-strand wire, which can be fitted to the fuse holder unit and allow pupils to make the connection to the bulbs easily.

Pupils connect a two-pole, two-way relay unit in place of lamp 2 and discover that the relay energises when light is incident on the cell. The low resistance of the cell under illuminated condition allows sufficient current to flow in the circuit to energise the relay.

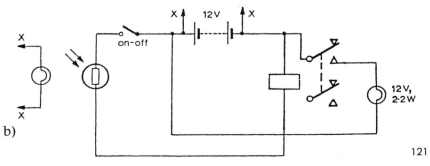

b)

With the light-source unit placed 10 cm from the photocell unit (measured by a metre rule placed alongside the equipment), the lamp (diagram (a)) in series with the cell glows brightly. As the distance between the light source and photocell is increased, the brightness of the series lamp falls, showing that the circuit resistance increases. When the beam of light is shut off from the cell, by placing a hand in the light beam, lamp 2 is extinguished, showing an even larger rise in circuit resistance. With the end of the light-source tube in contact with the end of the photocell-unit tube, the resistance of the cell is in the region 20–50Ω. In complete darkness (say when a hand is placed directly across the end of the photocell unit) the resistance rises to a value in excess of 500 000Ω (500 kΩ).

Some pupils may wish to include an on-off switch at this stage; others may also wish to include a lamp indicator unit to establish the condition of the relay, and should be encouraged to do so.

For relay operation, the maximum reliable distance between the photocell and the light source is approximately 0.5 m. Longer distances are often possible, but this depends to a large extent on the characteristics of the cell and relay. In later assignments a transistor/photocell arrangement allows a spacing of 3–4 m. Stopping a vehicle when the light beam is broken should present little difficulty for students.

Notes:

1 The circuits recommended in the assignment sheets are designed to obviate any serious damage to the photocell. However, if a fully illuminated cell is fitted directly across a supply, or a piece of equipment with only a low resistance is placed in series with the cell, the cell will overheat and be irrevocably damaged.

2 The ORP12 has a maximum power dissipation of 250 mW. Assuming at a maximum illumination (light-source unit and photocell unit in contact end to end) the resistance of the cell is 10Ω (though it is likely to be a little higher than this), the power dissipated in the cell when placed directly across a 12 V d.c. supply can be calculated as follows:

power dissipated in cell $= \dfrac{V^2}{R}$ watts (or I^2R or VI watts)

$$= \frac{12 \times 12}{10} \text{watts}$$
$$= 14.4 \text{ watts}$$
$$= 14400 \text{ mW}$$

The heat dissipated will burn out the cell.
The minimum series resistance permissible for 250 mW dissipation in the cell can be calculated similarly.

At a cell resistance of 10Ω, the maximum current for 250 mW dissipation can be calculated:

power $= I^2$ watts

$$\therefore \quad \frac{250}{1000} = I^2 \times 10$$

$$\therefore \quad I^2 = \frac{250}{10\ 000}$$

$$\therefore \quad I = 0.16 \text{ amperes (approx)}$$
$$= 160 \text{ mA}$$

Thus at maximum illumination (10Ω resistance) the maximum current through the cell is 160 mA. In order to limit the current to this value, the total circuit resistance can be calculated as follows:

Assuming the supply voltage is 12 V d.c.

$$R = \frac{V}{I}$$

$$\therefore \quad R = \frac{12}{160/1000} \ \Omega \ \left(160 \text{ mA} = \frac{160A}{1000} \right)$$

$$R = \frac{12\ 000}{160} \ \Omega$$

$$R = 75\Omega$$

The circuit resistance when using a photocell must never fall below this value. Since the cell resistance can be as low as 10 Ω at maximum illumination, a further 65 Ω (75–10), at least, must be provided by a device placed in series with the cell.

In the *Electrical switching assignments*, the following devices are used in series with a photocell:

) single-relay unit . . . coil resistance 185Ω
) bistable relay unit . . . coil resistance 185Ω

c) lamp indicator unit — 6 V, 0.36 W bulb . . . hot-filament resistance 100 Ω approx.

These figures show that no trouble should be experienced with relay units. With a series lamp unit, however, pupils should be discouraged from placing the light source and photocell units end to end, since the cell is then dissipating the rated 250 mW (approximately). Some normally rated 12 V d.c. power supplies will deliver more than 12 V, and thus the power dissipated in the cell may exceed the maximum permissible.* For the above reasons, pupils are asked to place the light source 10 cm from the photocell and to note that the effect when this distance is *increased*.

3   A relay/photocell oscillator can be simply constructed by making the contacts of a relay unit switch the light source on and off.

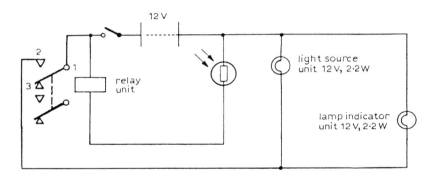

When the supply is switched on, the light source illuminates through the relay contacts 1 and 2. The photocell resistance falls and the relay energises. On energising the relay, contact 1 connects to 3, extinguishing the light source. The relay therefor is de-energised. Contacts 1 and 2 close again, switching on the light source. This sequence continues indefinitely. The lamp indicator flashes on and off to show the oscillation. A capacito fitted across the relay coil slows down the oscillation.

Teachers may wish to demonstrate the arrangement and/or set it as a buffer experiment.

---

*The variation of filament resistance with temperature further complicates the issue. At low brightness the filament resistance is less than 100 Ω.

Apparatus Required (15 pupils, 5 groups of 3)

five photocells units
five sensitive lamp indicator units (6 V, 0.36 W bulb) — if pupils use a second unit for comparison purposes, then five 12 V, 2.2 W units will also be required.
five light-source units (12 V, 2.2. W bulb)
five power supplies — 12 V d.c., 2 A — with safety fuse or trip
stackable 4 mm plug leads — assorted lengths
five 1-metre rules
five two-pole change-over relay units
five vehicles fitted with motor — made in *Electrical switching assignment 1*

Demonstration

Some teachers may wish to show the relay/photocell oscillator to the whole group at this stage.

Buffer Experiment

The relay/photocell oscillator assignment could be undertaken by a capable group: 'Connect a relay, photocell, and light source in such a way that when light strikes the photocell the relay energises, and at the same time the light source is switched off. To assist you in deciding what is happening, arrange for the relay to switch a lamp indicator on and off.'

Homework Questions

1 Fill in the blanks:

   i)   When light strikes a light-dependent resistor (photocell), the cell resistance .... When no light strikes the cell the resistance ....

   ii) A photocell is unsuitable for use in a .... atmosphere.

   iii) The inside of a photocell tube should be matt black to ....

2 Draw a circuit diagram to show how you would cut the supply to an electric motor when a beam of light is broken.

3 The circuit shown is intended to be used in an installation for checking lengths of metal bar placed between A and B. If the metal is short it must be rejected. Bars 2 m long or more are acceptable.

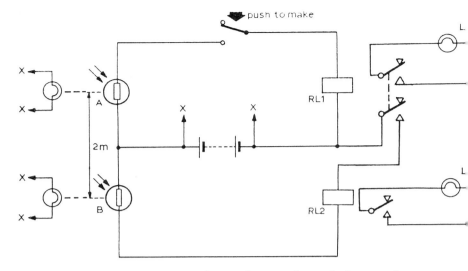

The operator presses the push-to-make switch to make a test.

i) Describe in detail how the circuit works.

ii) What is the purpose of L_1?

4 Draw circuit diagrams showing a light-dependent resistor (photocell) used as:

i) a switch, and

ii) a variable resistance.

5 A photocell and light source can be used in conjunction with an electromagnetic counter to find the speed of a motor shaft. Describe how you would do this. Include a circuit diagram, and explain how it works. (If you do not understand how an electro magnetic counter is constructed, you might ask to see inside one.)

Electrical Switching Assignment 8

This assignment should be attempted by more able pupils, but, alternatively, it could be demonstrated to other groups.

Content

Pupils are asked to make their vehicle oscillate in and out of a light beam, using a relay unit. This assignment should present little difficulty, since the reversing circuit has been used several times in previous assignments. However, when a capacitor is fitted across the relay coil — 2500 µF or so in the first instance — a number of possibilities arise. The circuit chosen by pupils will probably be:

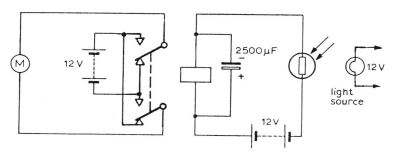

As the vehicle approaches the beam, the relay is energised since light from the source is incident on the photocell. When the beam is broken, the relay remains energised because of the capacitor. The vehicle, therefore, continues moving in the same direction. What follows depends largely on the design of the vehicle, the speed of the vehicle, its shape, and the value of the capacitor. This is fully explained in the follow-up.

To summarise, the possibilities are:

1   If the vehicle moves fairly quickly, it will carry on through the beam until light again strikes the cell, and there will therefore be no apparent effect — the capacitor has not had time to discharge sufficiently.

2   Even if the vehicle is slow, the light may be broken intermittently, and the capacitor charges each time light hits the cell. Again there is no apparent effect, the vehicle continuing through the beam and beyond.

3 If the capacitor discharges through the relay coil by a sufficient amount while the light beam is broken, the relay de-energises and reverses the vehicle.

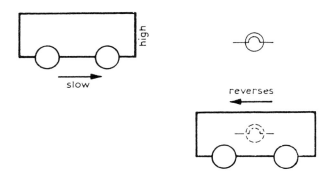

On leaving the beam, the relay (unexpectedly for the pupils) does not energise immediately. If the photocell were a perfect switch the relay would energise very quickly, but the finite resistance of the illuminated cell, say 100 Ω, delays the charge time and, therefore, the time at which the vehicle reverses. The vehicle may not reverse again until it is several centimetres from the beam.

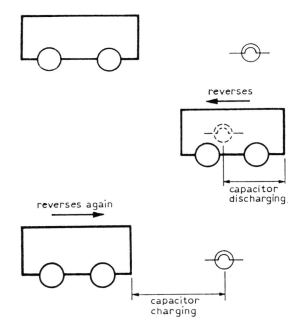

An Explanation for Delay Times

The following is intended primarily for the information of the teacher, though able pupils may benefit from a similar explanation.

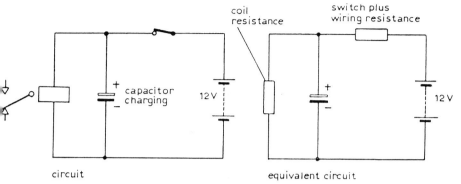

circuit   equivalent circuit

Assuming that the capacitor has become discharged, when the switch closes, the capacitor now charges quickly (passes a large current). This short-circuit effect prevents an appreciable current flow in the relay, which therefore cannot energise immediately.

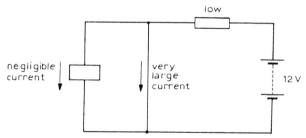

As a capacitor charges, its rate of charging diminishes.

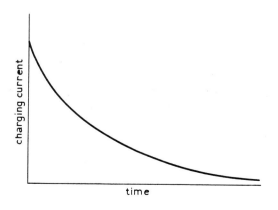

This represents a gradual increase in the effective d.c. resistance of the capacitor; the coil of the relay is therefore shunted by an increasing resistance. Since resistors in parallel share the total current in inverse proportion to their respective resistances, the current in the relay coil rises as that in the capacitor falls.

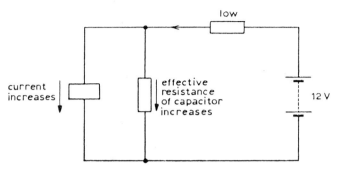

When the effective resistance of the capacitor reaches a certain high value, the current in the relay coil is sufficient to energise the relay. When fully charged, the capacitor ceases to absorb charge and in effect becomes an open-circuit (infinite resistance).

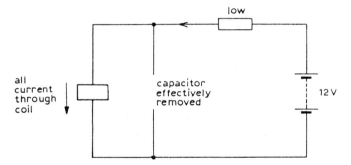

The capacitor stores a definite amount of charge from the 12 V supply, and the time it takes to do so is determined by the resistance in the charge path. If this resistance is small, the initial charging current can be large and the effective capacitor resistance rises rapidly. Since the resistance of a mechanical switch and its wiring is very low, the capacitor charges almost instantaneously and the relay 'appears' to close immediately.

When the switch is opened, the capacitor discharges through the relay coil.

Photocell Capacitor charging:

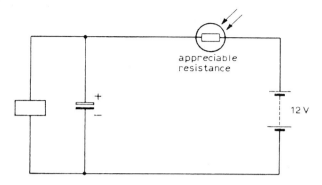

Assume the cell is not illuminated and the capacitor has discharged through the relay coil.

When light strikes the cell, its resistance falls but still remains appreciable; the capacitor initial charging current is therefore limited. Since the capacitor is uncharged, it represents a short-circuit across the relay coil.

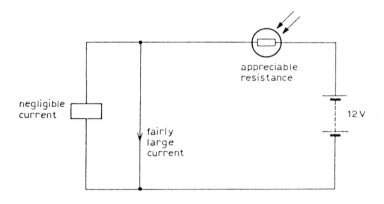

The resistance of the capacitor depends upon its state of charge and, since the charging current is limited, it remains a low resistance for an appreciable time, and there is a delay before the relay energises.

Photocell Capacitor discharging:

When light is removed from the photocell, the capacitor discharges through the relay coil.

131

Note: The rate of discharge also depends upon circuit resistance. If the terminals are short-circuited, the capacitor discharges almost instantaneously. However, the resistance in the discharge paths is the d.c. resistance of the relay coil i.e. 185Ω. Thus the initial discharge current is limited, and a considerable time elapses until the current falls to a value too low to maintain the relay in an energised condition.

Apparatus Required (15 pupils, 5 groups of 3)

five photocell units
five light-source units
five two-pole changeover relay units
five vehicles fitted with motors — as constructed in *Electrical switching assignment 1*
stackable 4 mm plug leads — assorted lengths
five 2000 µF, 25 volt capacitors
several assorted 25 volt capacitors, to enable groups to obtain approximately 8000 µF and 12 000 µF by parallel connection.
five power supplies — 12 V d.c. 2 A — fitted with fuse or trip

Demonstration

1   For the less-able pupils the whole of *assignment 8* should be demonstrated, or omitted completely.

2   For the above-average pupil, discuss the reason for vehicle behaviour as explained here.

Buffer Experiment

Set up the accompanying circuit and observe the effect of changing the value of *R* (0–600 ohms approximately).

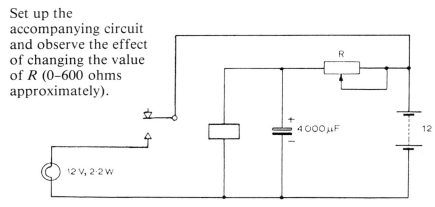

Homework Questions

1 Draw a circuit diagram which would enable your vehicle to oscillate in and out of a light beam. Describe the action of the arrangement.

2 The diagrams (a), (b), and (c) below show three similar circuits. The capacitors C in each are identical. The resistance of the resistor in (c) is twice that of the resistor in (b).

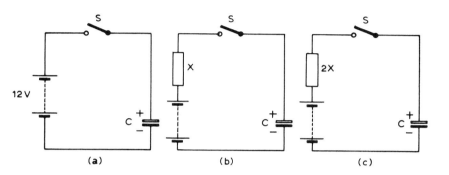

(a)   (b)   (c)

If the switches S are depressed simultaneously for a short time and then released, which of the capacitors is likely to be *storing*:

i) the most charge, and

ii) the least charge?

3 The diagrams (a), (b), and (c) show three similar capacitor-discharge circuits. The capacitors C are identical in value and are holding the same quantity of charge. The resistance, 2X, of the resistor in (c) is twice the value of the resistor in (b).

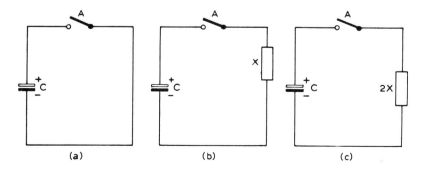

(a)   (b)   (c)

If the switches A are depressed simultaneously for a short time and then released, which of the capacitors is likely to have *lost*:

    i)   the most charge, and

    ii) the least charge?

4   Complete the following:
    When the supply is disconnected from a relay coil, the relay
    .... unless a .... is wired in parallel with the coil. The
    inclusion of this parallel .... keeps the relay .... for
    a time which depends upon the value of the .... the
    amount of .... it holds and the .... of the relay coil.

5   Describe and explain what happens when the light beam to the
    photocell in the circuit shown below is broken:

    i)   for a short time,

    ii) for a long time.

If a capacitor of about 2000 μF were fitted across the relay coil, how would the circuit now behave?

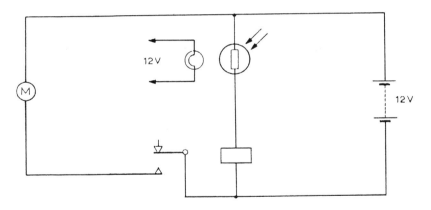

Electrical Switching Assignment 9

This assignment should perhaps be demonstratted by the teacher to all but the most able groups of children. At the end of *Electrical switching follow-up 8*, the suggestion was made of using a second relay to 'invert' the signal from a photocell. An illuminated photocell causes a series relay to be energised; in *assignment 8* this is a limitation if an immediate vehicle reversal, followed by a delay, is required. A capacitor fitted across a relay delays de-energising appreciably, but not energising. In order to make the vehicle reverse out of the light immediately, and reverse over a short distance, the breaking of a light beam must *de-energise* one relay which in turn causes a second relay to *energise*. This *second* relay is delayed using a capacitor.

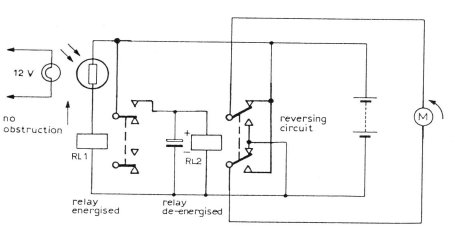

Immediately the vehicle breaks the light beam, RL1 de-energises. RL2 then immediately energises and reverses the vehicle motor via its reversing circuit. Once out of the light beam, RL1 again energises, removing the supply from RL2 (switching is now as shown in the diagram). RL2 remains energised for a while, however, owing to the capacitor discharging through the coil. When discharged, RL2 de-energises and the vehicle reverses towards the beam.

a) RL1 energised;
   RL2 de-energised.

b) RL1 de-energised;
   RL2 energised.

c) RL1 energised;
   RL2 energised via capacitor.

d) RL1 energised;
   RL2 de-energised
   (capacitor discharged).

Apparatus Required (15 pupils, 5 groups of 3)

ten-two-pole change-over relay units (if ten units are not available, some sharing will be necessary)

five photocell units

five light-source units

five 4000 μF, 25 V capacitors — any near value is satisfactory

five vehicles fitted with motor — as made in *Electrical switching assignment 1*

five power supplies — 12 V d.c., 2 A — fitted with fuse or trip

stackable 4 mm plug leads — assorted lengths

Demonstration

This assignment should be demonstrated to all but the most able pupils.

Buffer Experiments

1   Using a similar circuit, make the vehicle reverse indefinitely when a beam of light is broken.

2   If four-pole change-over relay units are available, construct a symmetrical-input relay bistable unit using two four-pole change-over relays. (Can you remember the circuit? Probably not, — but–before referring to your notes–try to set up the original circuit or any suitable alternative.)

Homework Questions

1   Describe the action of the following circuit on closing switch A:

   i)   when an obstruction is present, and;
   ii) when there is no obstruction between the light source and the photocell.

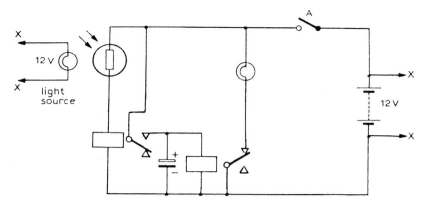

2   For the circuit below, describe what happens to lamp L when
     the switch B is operated, returned to the un-operated position,
     and operated again. Explain the behaviour of the lamp.

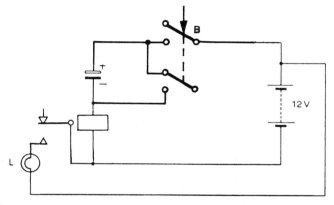

3   By means of suitable diagrams, show how a relay can be used:

     i)   as an inverter;

     ii) as a buffer.

4   A bell is required to ring approximately 10 seconds after a
     switch is pressed and to remain ringing for about 20 seconds,
     after which it must automatically stop. Draw a diagram of a
     circuit to perform this function, and explain how it works.

5   Describe what you would expect to happen in the circuit below
     if light strikes one or both of the photocells A and B. Give a
     reason for each of your answers.

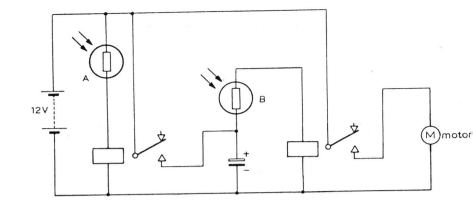

Electrical Switching Assignment 10

This assignment re-introduces the bistable unit, pupils again using it to control their vehicle, but in this instance photocells are used as switches in place of microswitches. This assignment should be demonstrated to all but the most able children. Some pupils might attempt the assignment in the normal way, but teachers should first ascertain, however, that sufficient apparatus is available, since the final circuit can be quite complex.

1   In the first part of the assignment the pupils connect a photocell across each pair of green socket on the bistable relay unit (signal input sockets) and not the effect when the light beams to the photocells are incident on the cells, and when the beams are broken. The purpose of these tests is to emphasise again that an illuminated photocell (which is usually the 'normal' photocell condition) behaves as a *closed* switch. The bistable unit does *not* behave as a bistable, however, since a requirement of the unit is that each green pair of sockets should receive a short-circuit only momentarily. Change-over cannot be achieved at the second pair of green sockets until the short-circuit has been removed from the first pair.

Using photocells, the bistable unit behaves as follows:

i)   Under 'normal' operation of a bistable relay unit *both* relays are never energised.

ii) By moving the obstruction in and out of the beam to relay A's circuit, relay A respectively de-energises and energises. Relay B is unaffected. Again this is 'abnormal'.

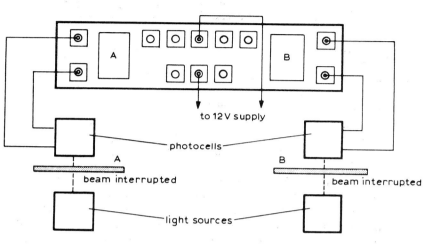

iii) The effect is similar to that in (b) but in reverse. The unit behaves 'abnormally'.

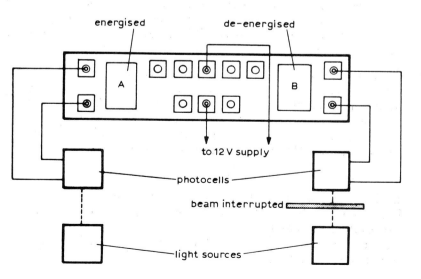

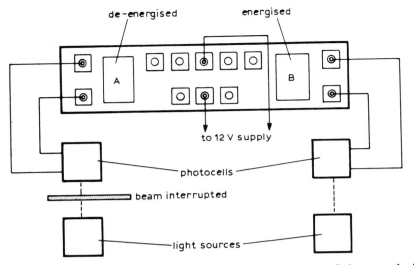

de-energised   energised

A   B

to 12 V supply

photocells

beam interrupted

light sources

iv) In situation (iv) the unit behaves 'normally'. If the supply is switched on with both obstructions in place, neither relay energises. If obstruction A is removed, relay A energises and remains energised whether the obstruction A is present or not. If obstruction B is removed (with obstruction A back in position) relay B energises and relay A de-energises. Relay B remains energised whether the obstruction B is present or not. By using the arrangement in this manner, the 'normal' condition for each photocell is the un-illuminated state (open-circuit switch). Short-duration short-circuits are applied across each pair of green sockets by removing the obstructions momentarily. This situation is analogous to the use of normally-open microswitches.

2   Once the reasoning above has been understood by pupils, they should quickly realise that if photocells are to be used in conjunction with a bistable unit, and if the photocells will normally be illuminated, a relay must be incorporated in each photocell circuit to *invert* the photocell signal. When each beam is now *broken* in turn, the bistable will operate correctly.

See diagram (a) on page 141

Both relays are normally energised (photocells illuminated — beams unbroken). Breaking a beam momentarily causes the associated relay to de-energise, momentarily placing a short-circuit across one bistable input (the other input at this time being open-circuited).

3   A vehicle can now be made to 'oscillate' between two beams of
    light, since the motor reverses direction each time a beam of
    light is broken. See diagram (b) below.

(a)

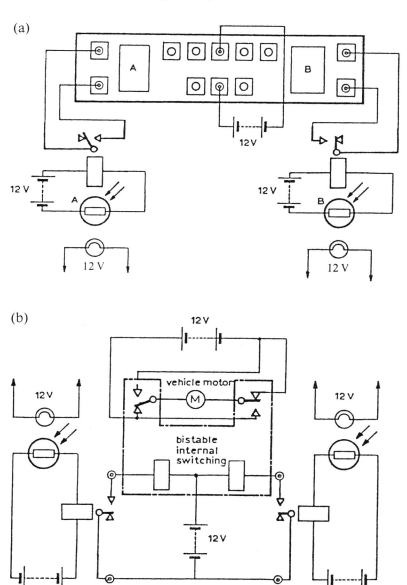

(b)

Notes

1   In the above circuits, more than one 12 V supply has been shown. In practice only one would be used, but pupils should be encouraged to show separate ones in any complex circuit diagram, as greater clarity can be achieved by doing so.

2   When pupils test the bistable unit in the first part of this assignment they should be encouraged to connect a pair of lamp indicator units to the bistable-unit external-switching sockets, to indicate the states of the relays.

Apparatus Required (15 pupils, 5 groups of 3)

five bistable relay unit (if ten units are not available, some sharing will be necessary)
ten two-pole change-over relay units
ten photocell units
ten light-source units
five vehicles fitted with motor — as made in *Electrical switching assignment 1*
five power supplies — 12 V d.c., 2A — fitted with safety fuse or trip stackable 4 mm plug leads assorted lengths
ten lamp indicator units

Demonstration

As has already been suggested, for all but the more able children, the whole of this assignment should be demonstrated. If the teacher wishes it, all could read through the assignment, and one group of children set up the equipment, taking into account suggestions from the rest of the pupils.

Buffer Experiments

1   The following circuit enables a bistable relay to operate from a single input. Set up the arrangement and test it. A single input bistable should change-over at the first input pulse and change back again at the next pulse. In this case the pulses are short-circuits provided by switch S.

How do you think the performance of the circuit is affected by:

i)   the length of the input pulse,
ii) the time between pulses?

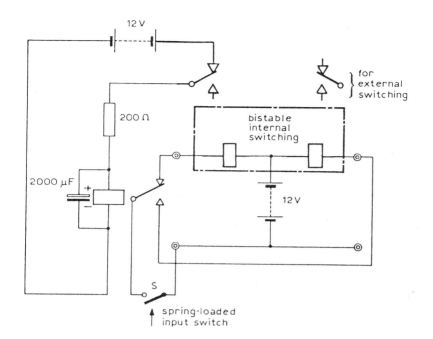

1 A photocell can be used in a switching capacity. Does such a cell behave as a 'closed' or as an 'open' switch:

i) when the cell is illuminated, and;
ii) when the cell is not illuminated?

Why is a photocell not quite as good a 'switch' as a mechanical type?

2 Describe and explain how you would set up a circuit consisting of bistable relays, a relay oscillator, an electromagnetic counter, and any other apparatus you need, in order to time how long it takes to walk or run 100 m.

3 A traffic-light system, consisting of red and green lights only, is often used when road works prevent the use of both sides of the road. Devise a suitable circuit to control a set of lights for this purpose, and describe how it functions.

4 A burglar alarm is required for your school. Design a suitable system and explain how it works.

143

Linear Motion

General Introduction

It is recommended that the pupils tackle this section after completing the mini-projects which follow the *Electrical switching assignments*.

During the course so far, pupils have been concerned primarily with rotary motion, but in future project work linear motions may be required for a number of purposes — e.g. feeding, spacing, rejecting, steering, clamping mechanisms, etc. — and these assignments are designed to make pupils *aware* of the various methods available and to enable them to *select* the most suitable system for a given design requirement.

Pupils are asked (in *assignment 1*) to think of as many ways as possible of converting rotary motion to linear motion and then to select one of their ideas, construct a Meccano version, and test it. Fortunately, the range of Meccano components enables most of the commonly used mechanisms to be assembled with little difficulty. A wide range of solutions should be encouraged, to enable the properties of the various systems to be compared.

However, it is unlikely that all the linear-motion problems to be met in later project work can be solved satisfactorily by using an electric motor and a standard mechanism of some kind; for instance, if a large force is to be applied then a very robust mechanism is essential, which might well cause constructional problems and operate unsatisfactorily. However, such a force can be provided conveniently and efficiently by means of a pneumatic (or hydraulic) cylinder; whilst if a short fast movement is required this might be better produced by means of a solenoid. It is therefore considered desirable that pupils should have the opportunity of using components such as solenoids and pneumatic cylinders to enable efficient and realistic solutions to be produced.

Pressure Units

Pressures used to be measured in lbf/in^2. Now the correct unit for pressure is the pascal (Pa). One pascal is one newton per square metre. The pascal is a very small unit of pressure. Common practice in the pneumatic industry is to measure pressure in bars (1 bar = 100 000 Pa).

It is quite likely that you will find pressure gauges calibrated in kgf/cm² or bars.

The relationship of these units is as follows.

1 pascal	= 0.00001 bar
1 standard atmosphere	≏ 1.01 bar
1 kgf/cm²	= 0.98 bar
1 lbf/in²	= 0.07 bar

It will be noted that the bar and kgf/cm² are both approximately equal (the bar being exactly 100 kPa) and so conversion of gauges calibrated in these units is comparatively straightforward.

Approximate pressure equivalents:

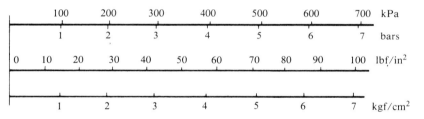

Cylinders are now made with metric diameter bores. However schools may still have imperial equivalents. The following approximate equivalents will prove satisfactory for the linear motion investigations.

½"	=	12 mm dia.
¾"	=	20 mm dia.
1"	=	24 mm dia.

Assignments 2 and 3 investigate the properties of solenoids and pneumatic cylinders and have two main aims. Firstly to discover the operating characteristics of both the solenoid and the pneumatic cylinder. This knowledge, together with that gained from the mechanisms discussed in *assignment 1*, should enable pupils to select a suitable system for a particular application.

Experience has shown that *assignment 2* is more satisfactorily dealt with as a teacher demonstration activity to emphasize these aims and demonstrate use of the associated equipment which is new to the pupils.

Secondly, *assignments 3.1–3.7* provide suitable training and experience in the investigational approach. Since this is the first time that pupils have been required to carry out quite difficult investigations, considerable guidance is essential if results are to be satisfactory. Although the assignment should ensure that each investigation is tackled in methodical steps, it is advisable for each pupil to study the notes entitled *The approach to investigational experiments*, before attempting any of *assignments 3.1-3.7*. In most of the investigations, pupils are asked to draw a graph from the results obtained. Most pupils are able to draw reasonably good graphs, but very few are able to interpret them and draw suitable conclusions. For this reason, pupils should refer to *The use of graphs*, before the investigations are attempted. It is not intended that pupils should 'learn' the contents at this stage, but they should read through the notes in order to be aware of the essential points.

As a result of carrying out these investigations, the pupils should:

a) become more methodical in their work;
b) improve their powers of observation;
c) learn to use and read instruments more accurately;
d) learn to take more than one reading to ensure suitable accuracy;
e) record results in a clear tabulated fashion;
f) draw graphs from their results if necessary;
g) produce suitable conclusions from analysing the results — this is particularly important.

Thorough training in these techniques should help the pupils in future investigational work, in analysing ideas for project work, and in everyday-life situations.

For *assignment 2*, commercial types of solenoids should be available for the pupils to examine.

For *assignments 3.1-3.3* the special multi-tapped solenoid is recommended. The power supplies and switch units used in the *Electrical switching assignments* are suitable for use with the solenoids.

The provision and use pneumatic cylinders and associated equipment might present some initial problems. However, it is considered important that pneumatic equipment is made available for the following reasons.

1 Comparatively large forces are easily obtained, e.g. a 25 mm

diameter cylinder produces at a working pressure of say 6 bar a force of approximately 250 N — a typical feed force for drilling a 4 mm diameter hole in a piece of mild steel.

2 Pupils are made aware of the industrial application of the equipment, and can solve relevant problems in a realistic way.

3 Pupils are provided with a wider range of techniques with which to solve problems.

4 In principle, pneumatic components are very similar to hydraulic components, and references can be made to the use and advantages of hydraulic systems. However, in using pneumatic components one avoids the danger of the high forces produced by hydraulic systems and the possibility of burst pipes and leaks. With the pneumatic system, used air escapes to the atmosphere, whereas all used hydraulic fluid must be returned to the pump.

5 With the addition of fluidic sensors, which provide very sensitive detecting facilities (analogous to photocells etc.), pneumatic systems can be used in a wide range of situations.

6 Logical circuitry can also be applied to pneumatic systems.

Since pneumatic valves are analogous to electrical switches, pupils having completed the *Electrical switching assignments* experience little difficulty in using the new equipment.

Although pneumatic equipment is rather expensive, the components are quickly taken apart and used again and again. They are very robust in construction and should give many years of reliable service.

Safety

Attention should be drawn to the obvious dangers of trapping parts of the body between mechanisms operated by compressed air. Pupils should be warned not to investigate an apparently malfunctioning circuit by pressing valves, located at the end of a piston travel, with their fingers. For example a common fault is for pupils to transpose the pilot line connections to a five-port valve. Pressing the three-port valve at one end of the piston travel then causes the piston to move out and trap their finger. Use of a flat piece of metal held in the hand avoids this danger. Attention should be drawn to three further hazards.

1   Very serious bodily damage can arise if compressed air is misused, e.g. an open pipe directed into a ear etc.

2   Long pipes, when disconnected at one end, swing about dangerously when carrying high-pressure air.

3   Warn pupils, by demonstration if appropriate, of the potentially dangerous forces operating in cylinders of large diameter.

The pressure reducing valve should be the first component in the supply line from the compressor/reservoir. It incorporates a filter and probably has only two available connections, an inlet and an outlet port. Reference should be made to the Equipment Guide regarding insertion of screwed connections in the valve body.

The dangers mentioned above can be minimised by introducing the following circuit into the main supply.

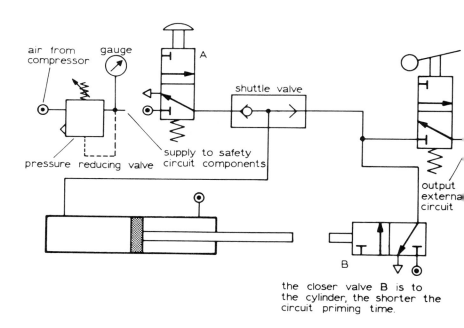

the closer valve B is to the cylinder, the shorter the circuit priming time.

To activate the circuit, push-button A is pressed, which, via the shuttle valve, provides primary air supply to the push side of the cylinder and, as discovered in *Linear motion assignment 3.6*, equal pressure on each side of the piston of a double acting cylinder causes the piston rod to move out slowly.

When the piston rod activates valve B, air is supplied to the external circuitry and also to the cylinder via the shuttle valve. Valve A may then be released because, if there is sufficient back pressure from the external circuitry, the piston rod will continue to activate valve B. However, if the outlet pipe is disconnected there will be insufficient pressure applied to the push side of the piston and the piston rod will retract, releasing valve B and therefore cutting off the air supply. Also if pipes are disconnected from the external circuitry with the pressure *on*, the supply will be cut off. To restore the supply, pipes must be re-connected, valve A held down until the piston rod again activates valve B, then valve A released.

Admittedly, air pressure is supplied to the external circuitry whether pipes are connected or not if valve A is held down, but on releasing this valve the supply is cut off and this should provide adequate protection.

By using a 25 mm dia. cylinder a difference of approximately 0.3 bar between supply pressure and output pressure is detected, which provides protection against disconnected pipes but does not cause the supply to be cut off under normal working conditions. Because the circuit detects differences between supply and output pressures, it will operate at all but the very lowest supply pressures.

The most satisfactory means of obtaining compressed air is by means of an electrically driven compressor incorporating a storage tank. (The hydrovane compressor does not require a tank; it is therefore much lighter, but unfortunately much more expensive.) A capacity of at least 0.1 m³ is required and an operating pressure in excess of 3 bar (normal working pressure is usually 6 bar). It is strongly recommended that a compressor with automatic pressure control is obtained.

To cut costs, a second-hand compressor might be considered, or the assembly of separate components, e.g. a compressor driven by a cheap second-hand electric motor with the addition of a pressure switch, storage tank, etc. This will be a satisfactory solution only if the equipment is reliable, otherwise a great deal of frustration will occur.

Some schools with good workshop facilities may be tempted to make some pneumatic components (cylinders etc.) to reduce costs, but this is not recommended as it is unlikely that such equipment would prove satisfactory. Absolute reliability of all components used in the course is considered essential if the pupils are to gain the confidence necessary to achieve success in the more complex project work.

The hiring of bottles of compressed air (**NEVER OXYGEN** because of the oil present in the valves and cylinders) is another possibility. If a suitable regulator and gauge are available, this may be a satisfactory solution in the short term.

It is considered essential throughout this course that pupils handle the equipment, investigate possible applications, and also used it to solve the problems confronting them in later project work.

Unless a source of compressed air and a range of valves and cylinders are available for practical work, it is strongly recommended that the section involving pneumatic equipment is omitted it is debatable whether one should talk about equipment unless it is available for the pupils to handle, investigate, and use.

A compressor, together with a range of pneumatic components, might be shared by a number of schools, each school using the equipment for a half-term or so. Initially, this may seem to be a satisfactory solution, as it would allow pupils in each school to cover the basic work in pneumatics, but the pupils would be deprived of the equipment when tackling their own project problems. Unless the pneumatic equipment is available during the whole period of the project (constant experimentation and testing is usually required once the initial ideas have been formulated), one should question, perhaps, whether to undertake the assignments involving the use of pneumatic components.

Choice of Pneumatic Components

There are manufacturers of pneumatic components, several of whom make components suitable for use in this course. Factors to be considered include the following.

1 *Cost*

   Expensive, well-made components are generally more reliable than cheaper versions; but, as all are guaranteed to operate for a million or more cycles of operation, the cheaper valves are satisfactory for school use.

Size

If the pneumatic components are to be used in conjunction with Meccano, or in the construction of working devices of limited size, the components themselves must be small. 12 mm (½") cylinders have been used successfully, and even smaller ones are available. Valves are sometimes difficult to accommodate, but a microswitch operating a solenoid valve could be used, or alternatively the sensing can be carried out by fluidic devices.

Shape

It is essential that the components can be easily fixed into position, and those having several fixing holes are more satisfactory.

Screwthreads

Unfortunately, components supplied by different manufacturers may not be compatible, because a variety of screwthreads are used. Adaptors can be obtained to make all connections compatible; these add to the initial expense, but do prevent improperly connected fittings from damaging the ports and rendering an item useless. A damaged adaptor can be replaced cheaply.

Pipes

Nylon pipes are essential, and several types of fitting are available. All methods are quite satisfactory, but it must be remembered that, during this course, pipes will be connected and disconnected far more frequently than during industrial applications. A pipe fitting which protects the end of the pipe will probably prove most satisfactory.

A supplier of specialised equipment for this course is:

Economatics Education Division Ltd
Epic House
9 Orgreave Road
Handsworth
Sheffield S13 9LQ

Linear Motion Assignment 1

In this assignment, pupils are asked to devise means of transforming rotary motion to linear motion. All pupils should be able to suggest at least three methods, and some may produce as many as six or seven. It is advisable to introduce this assignment with a class discussion, asking pupils to give known examples of rotary-to-linear motion, and for the teacher to assist in identifying them under the categories referred to in the follow-up book. Class experience can be maximized by arranging for each sub-group to construct a different type of mechanism. Some pupils may encounter difficulty in putting ideas on paper because their diagrams become unnecessarily complicated.

It may be advisable to give some hints on how to represent various components to enable simple line diagrams to be made, for example:

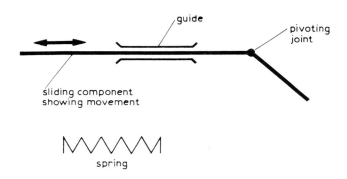

It is important that as many different ideas as possible are developed; this will provide training for later project work when alternative methods of solving a problem must be explored. To achieve this, it is essential that a certain time (20–30 minutes) is given to *section 1*, during which time no pupil is allowed to proceed with *section 2*. Without such a restriction, some pupils, eager to build a model, will be satisfied with one or two obvious solutions and will not give the problem sufficient thought.

When the teacher is satisfied that *section 1* has been satisfactorily completed, he may wish to ensure that a number of different solutions to the problem are constructed. If two groups wish to construct the same system, one of them could be encouraged to be more ambitious and to attempt a different solution. Experience has shown that it is possible for five alternative solutions to be produced with little or no direction from the teacher.

The results obtained from *sections 3, 4,* and *5,* depend upon the ability range of the pupils; the teacher may consider that *sections 1* and *2* provide sufficient involvement and challenge for his or her pupils.

When the construction work and tests have been completed, it is suggested that each group demonstrates its own version, giving details of constructional problems, advantages and disadvantages, etc. This approach not only provides a great deal of information, but it gives the pupils an opportunity to talk about their work. The pupil with limited verbal ability may have useful information about a mechanism he has made.

Whilst some of the facts in the follow-up will have been dealt with, it is suggested that the teacher and the class should look through the notes together, so that further explanations can be offered:

A few notes from the follow-up may be desirable, but these should be kept to a minimum — the aim of this assignment is to make pupils *aware* of the various systems, to help guide their choice of mechanisms in the future.

Apparatus Required (15 pupils, 5 groups of 3)

five motors

assorted Meccano angle, strip, angle brackets, gears, wheels, nuts and bolts, etc.

additional Meccano items:

rack strips	(part no. 110 or 110a)
slide pieces	(no. 50)
eccentrics	(no. 130 or 130a)
fork pieces	(no. 116 or 116a)
slotted strips	(no. 55 or 55a)
corner brackets	(no. 133)

five 1 kg weights

With the above equipment, the following mechanisms can be constructed:

a) crank and slider;
b) peg and slot;
c) eccentric;
d) rack and pinion.

Demonstrations

1 If a particular mechanism has not been constructed and the teacher feels that the pupils should be able to see exactly how it operates, e.g. the peg-and-slot mechanism, a model should be made and demonstrated later.

2 It is very important that, having constructed Meccano models, pupils should also be shown how the mechanisms are used in engineering. It should be possible to find examples for demonstration purposes when the follow-up is being studied

Examples which should be readily available include:

a) *Crank and slider*
   Car or motor-cycle piston, gudgeon pin, connecting rod, and crank-shaft
b) *Peg and slot*
   Shaping machine
c) *Cam and follower*
   Car or motor-cyle valve gear
d) *Eccentric*
   Model-steam-engine valve gear; some types of paper punch
e) *Rack and pinion*
   Hand-operated carriage movement on lathe; feed mechanism of bench drilling machine; car steering box
f) *Screwthread*
   Vice; car-jack; lathe leadscrew

Buffer Experiments

1 In the screwthread and rack-and-opinion systems, the rotation must be reversed to change the direction of movement. Using such an arrangement, driven by an electric motor, devise a method of making the moving part reciprocate backwards and forwards continuously.

2 Provide a screw-type car-jack. What is its velocity ratio?

Homework Questions

1 Velocity ratio $= \dfrac{\text{distance moved by effort}}{\text{distance moved by load}}$

   Mechanical advantage $= \dfrac{\text{load}}{\text{effort}}$

$$\text{Efficiency} = \frac{MA}{VR} \times 100\%$$

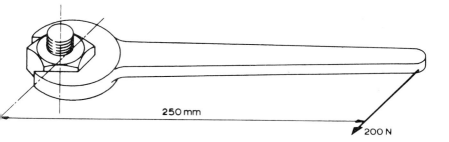

250 mm

200 N

In this example the screwthread has a 'lead' of 1 mm, i.e. in one revolution the nut moves 1 mm along the bolt.

If the efficiency of the 'system' is 15%, calculate the force produced.

In a four-stroke petrol engine, each valve opens once for every four strokes of the piston. Make a simple sketch of the main components of the engine, showing the valve-operating mechanism.

Compare the cam and eccentric mechanisms. What are the advantage and disadvantages of each?

When the steering wheel of a car is turned, the rotary motion is converted to linear motion by the steering box. Make a sketch of the machanism you would prefer to use for a steering box.

From the graph shown below:

i)   What is the minimum distance that the ram projects from the machine?

ii) What is the maximum project of the ram? How many seconds are required to reach this position?

iii) What can you say about the speed of the ram between A and B and between B and C?

iv) Between what times is the ram moving at its greatest speed? Calculate this maximum speed.

v)   At what time is the ram projecting 50 mm from the machine on its return journey?

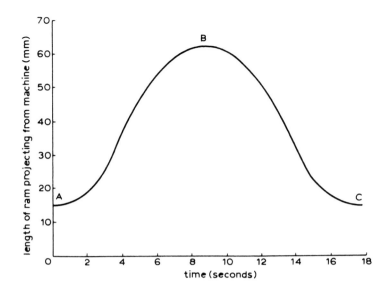

6   Account for the differences in the shape of the two graphs
     shown below.

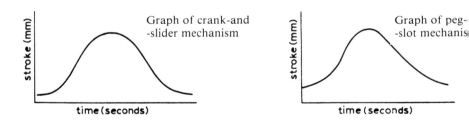

Linear Motion Assignment 2

This assignment introduces the pupils to solenoids and pneumatic
cylinders, the properties of which will be thoroughly investigated in
Linear motion 3.1–3.7.

Pupils are asked to examine the construction of various solenoids
and pneumatic cylinders provided, and to operate each through the
appropriate power supply in order to establish basic properties.
Some groups may wish to carry out some simple testing at this

stage, and this should be encouraged, provided that the tests are simple ones. However it is perhaps appropriate to ensure that all pupils are able to observe a teacher demonstration of the pneumatic equipment at this stage.

For this assignment, and for *Linear motion 3.1-3.7*, it is unlikely that sufficient apparatus will be available to allow five groups to attempt the same investigation. It is suggested that three sets of pneumatic equipment and three sets of electrical equipment are provided, and that the class is divided into the usual five groups. (The spare set should ensure that pupils waste little time in waiting for apparatus.)

Some introduction must be given to the source of compressed air, whichever system is used, and to the control of the air through the pressure regulating valve and on-off valve.

The following layout is recommended:

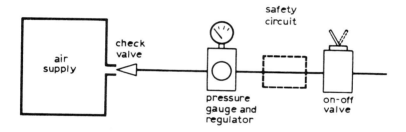

The operation of the compressor, the need for a storage tank and safety valve, and the method of maintaining a reasonably constant-pressure supply through pressure-sensitive electrical switches etc. should be outlined at this time. The operation of the pressure regulator should be demonstrated: there is a need, when reducing pressures, to reduce further than necessary and then to increase to the required value.

Stress that airtight joints can be made without excessive tightening of components; if too great a force is applied the fittings may be damaged, or indeed broken.

It may be considered desirable to provide further information on the construction of a pneumatic cylinder, e.g. the position, shape, and material of the airseals. Since the construction will vary, for specific details consult the manufacturers' catalogues. However, it is felt that the principles of construction are more important at this stage than individual design features.

Apparatus

commercial solenoids of different sizes and types

12 V d.c. power-suply units

on-off switch boxes

leads

single-acting pneumatic cylinders of varous diameters and strokes

double-acting pneumatic cylinders of various diameters and strokes connecting pipes

a source of compressed air with regulator and gauge, with three on-off valves for take-off points.

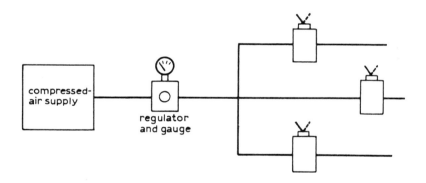

Demonstrations

1   Demonstrate and explain the operation of the pneumatic equipment.

2   Demonstrate the use of the regulator — if reducing the pressure, reduce below the value required, and then build up the pressure.

3   Demonstrate the type of pipe fittings used, emphasising that over-tightening may cause damage or breakage. The method of making up a pipe could also be demonstrated if necessary.

4   Demonstrate various applications illustrating the use of

solenoids and pneumatic cylinders, e.g. a relay, an electric bell, a vehicle trafficator, a pneumatic door control.

Films dealing with the use of pneumatic equipment in industry are available from the manufacturers of pneumatic equipment.

Buffer Experiments

Replace the solenoid armature with:

i) a piece of brass rod of the same diameter,

ii) a piece of alluminium rod of the same diameter,

iii) a piece of steel of approximately half the diameter of the original armature.

In each case record your observations and, if possible, give an explanation for the results.

Operate a pneumatic cylinder at various air-supply pressures. Estimate the effect the change in air pressure has on the force produced.

Homework Questions

Produce sketches showing some applications of solenoids and pneumatic cylinders with which you are familiar.

Compare the movements produced by:

i) the armature of a solenoid,

ii) the piston rod of a single-acting pneumatic cylinder.

iii) the piston rod of a double-acting pneumatic cylinder.

Linear Motion Investigations 3.1–3.7

Refer the pupils to the notes on *The approach to investigational experiments* and *The use of graphs* before any of these assignments are attempted.

To be completed successfully, these investigations demand skill in setting up the apparatus, accurate observation and measurement, thorough and orderly recording of results, and the ability to reach a reasoned conclusion.

Pupils should be encouraged to:

a) check their results by taking more than one reading (averaging the readings if necessary);
b) tabulate the results, if appropriate;
c) record accurately what they observe, whether it appears 'correct or not;
d) draw graphs to establish relationships;
e) develop meaningful conclusions.

Experience has shown that initially pupils find these investigations difficult, but they make good progress once familiar with the apparatus and the techniques involved.

It may be advisable to partially demonstrate, *investigations 3.1* (solenoid) and *3.4* (pneumatic cylinder) before the pupils attempt any assignment, to ensure that the basic essentials are fully appreciated from the start.

When working the solenoid investigations it is important that pupil appreciate that the possible variables are:

a) number of turns;
b) penetration of the armature into the coil;
c) the current flowing through the coil.

A common cause of confusion is for pupils to believe that if they maintain a constant voltage across the solenoid that the current wil remain constant, which of course it does not if the number of turns is changed.

Pupils should be allowed to work through these investigations at their own rate; some will finish all seven (and the buffer experiments) during the time others take to complete three or four. Providing all pupils have successfully tackled a number of investigations, and have developed the necessary techniques, it is suggested that the teacher demonstrates the remaining investigations, to prevent holding up the faster workers. A general discussion of results should follow, and some general conclusions concerning the properties of solenoids and pneumatic cylinders should be developed and recorded.

The teacher should repeat *investigation 3.1*, this time using a solenoid with a steel insert (bolt). Whilst this investigation may have been attempted by some of the faster workers (as a buffer experiment), it should now be demonstrated to all, and the effects discussed in detail. This is the form of solenoid that the pupils will invariably use in later projects.

The follow-up should provide all the information required, although results may vary slightly from those indicated. Since the apparatus used is relatively crude, it is wise not to interpret the results too rigidly.

It should be remembered that the main purpose of these assignments is that:

a) pupils appreciate certain properties of solenoids and pneumatic cylinders;

b) pupils develop the skills required to carry out investigational work which in future may not be so clearly directed.

Apparatus

Solenoid Investigations 3.1–3.3

Apparatus required per group:
one multi-tapped solenoid unit
one graduated armature
one 0–100 gram (0–1N) spring-balance
one retort stand with two clamps
one switch box
one 0–2A ammeter
five leads
one variable-voltage power supply

NOTE: Do not attempt to control the current flow through the solenoid using the variable resistors from the course equipment, as they will overheat.

Pneumatic cylinder investigations 3.4–3.7

Apparatus required per group:

either a) 12 mm dia. double-acting cylinder with 0–100N spring-balance and frame mounted on a board
or   b) 20 mm dia. double-acting cylinder with 0–100N spring-balance and frame mounted on a board
or   c) 25 mm dia. double-acting cylinder with 0–1250N spring-balance and frame mouned on a board.

Note: The stroke of each cylinder must be adequate to extend the spring-balance over the working range.

Details of frame and mounting board

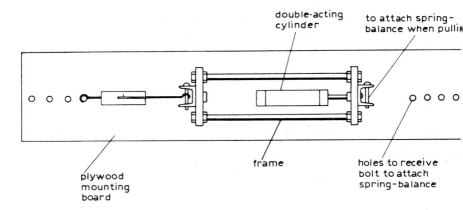

double-acting
cylinder

to attach spring-
balance when pullin

plywood
mounting
board

frame

holes to receive
bolt to attach
spring-balance

Dimensions will depend cylinders and spring-balances available.

one air-pressure supply
one regulator and gauge
one on-off valve
one pipe

Ideally each group should have its own regulator and pressure
gauge; but it is possible for three groups to work together from one
gauge, though each will need an on-off valve (as suggested for
Linear motion 2).

for *investigation 3.6*, each group will need a tee-junction and three
connecting pipes.

Demonstrations

1   Set up and partially demonstrate *investigations 3.1* and *3.4*
    before these investigations are attempted by pupils.

2   Some of the later investigations could be demonstrated, to
    enable the slower workers to catch up.

3   Repeat *investigation 3.1* using a solenoid fitted with a steel
    insert. Record and discuss the results.

Buffer Experiments

1   Repeat *investigation 3.1* using a solenoid fitted with a steel
    insert.

2   Using a cylinder of known piston diameter, measure the diameter of the piston rod and by experiment establish the frictional forces when (a) pushing, (b) pulling.

Homework Questions

1   A solenoid fails to produce an adequate pull force. Suggest how this can be increased (a) without altering the solenoid, (b) by re-designing the solenoid.

2   A double-acting pneumatic cylinder has a piston of 50 mm diameter. Draw a graph of pressure against force for this cylinder to enable you to read off the theoretical (ignoring friction) push and pull forces for all pressures up to 6 bar.

3   A pneumatic cylinder has an effective piston diameter of 40 mm. The air pressure applied is 7 bar. The average frictional force is 80 N. What force is exerted by the piston rod? (Take π to be 22/7.)

4   The armature of a relay constructed in the school workshops fails to move away from the coil when the latter is de-energised. What could be the reason for this? Suggest how the relay could be made to function correctly?

5   The diagram below shows part of a car braking system. What force is applied to each pad of the disc brake?

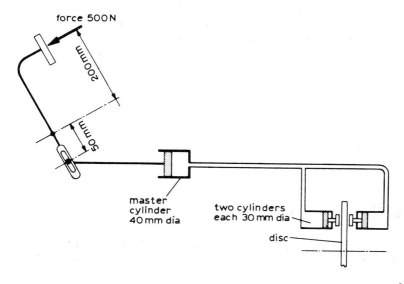

force 500 N

200 mm

50 mm

master cylinder 40 mm dia

two cylinders each 30 mm dia

disc

Pneumatic Control ————————————

General Introduction

In the *Linear motion assignments*, the pupils investigated the properties of pneumatic cylinders. Further knowledge and experience in the use of pneumatic-control equipment is now required.

To help pupils appreciate the similarity between pneumatic and electrical circuits, reference is made throughout these assignments, to the equivalent electrical components and electrical circuits used in the *Electrical switching assignments*.

Symbols

The symbols used in these assignments are the recommended standard symbols. Valve switching positions are described in the pupil's assignments book in the *Linear motion-pneumatic cylinders* section.

Equipment

For guidance on the compressed air supply and other equipment, refer to the teachers' notes on *Linear motion* and Equipment Guide page 67.

The use of a combined filter and regulator is recommended, but a lubricator is not considered essential since components are unlikely to need lubrication when used for relatively few operations. Dust entering the valve ports is more likely to cause unreliable operation.

Types of valve

Two types of valve are particularly suitable for the work undertaken in the course:

a) the poppet valve,
b) the spool valve.

The Poppet Valve

The poppet valve mechanism consists of a ball held against a seal by means of a spring; operation is achieved by holding the ball away from the seal. In this type of valve the air supply must assists the spring in holding the ball against the seal. Air can flow from the inlet to the cylinder connection only when the valve is operated, and from the cylinder to exhaust when unoperated.

The advantage of this valve over the spool type is that frictional resistance is minimal and therefore actuating forces are comparatively low, which makes it particularly suitable for use as a 'sensor', i.e. converting mechanical movement to air-pressure change.

In these assignments these valves are referred to as 'on–off' valves.

The Spool Valve

In this type of valve, a spool (basically a cylindrical rod with a number of annular grooves cut in it) is moved to control the flow of air. Because of the friction between the spool and the sealing rings, these valves require actuating forces two to three times greater than those for the poppet-type valves.

The function of the sports of the spool valve may be interchanged.

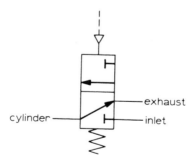

For example with the air supply connected to:

a) the inlet port — air pressure is directed to the cylinder when the valve is actuated (normal use);

b) the exhaust port — air pressure normally flows to the cylinder but is cut off when the valve is actuated (the inverse of (a));

c) the cylinder port — air presssure normally flows to the exhaust port, and to the inlet port when the valve is actuated.

165

The three-port valve shown is used in each of the above ways in *assignment 4*.

In these assignments, the three-port spool valve is referred to as a 'single change-over valve', and the five-port as a 'double change-over valve', to emphasise the fact that the pneumatic valves perform the same function as the electrical switches used in the *Electrical-switching assignments*.

Actuators

Both the 'poppet' and 'spool' valves are available with a wide range of actuators (see *assignment 1*).

Approximate operating forces/pressures:

Actuator	*Spool valves*	*Poppet valves*
Push button/plunger	40 N	12 N
Roller operated	20 N	6N
Lock-down lever	10 N	3 N
Double pilot	1.5 bar	—
Single pilot, spring return	3 bar	1 bar
Diaphragm, spring return	0.5 bar	0.2 bar

Working Pressures

It is recommended that normal working pressure be in the region of 4–6 bar — it must be in excess of 3 bar since this pressure is required to operate a pilot air-actuated, spring-return spool valve.

Note that if air stored in a reservoir is being used to operate a valve, as in *assignment 4*, it is recommended to use a diaphragm valve, for the following reason.

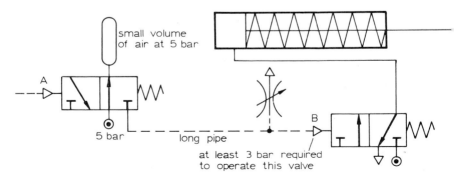

When valve A is operated, air from the reservoir at 5 bar flows through the pipe to valve B. With the increase in volume there is a corresponding fall in pressure. If the pipe is long there may be insufficient pressure to operate valve B. In this case use a diaphragm-actuated valve at B, which will operate at pressure as low as 0.2 bar (this point is made in *follow-up 4*).

Conversely, in the circuits in *assignment 5*, a high-pressure operated valve is required, e.g. a pilot-operated spring-return spool valve, to enable a reasonable time delay to be obtained.

Organisation

With limited pneumatic equipment, it is unlikely that all pupils can be involved in practical work at any one time; there is then a temptation for the teacher to merely demonstrate all the investigations. This should be avoided if the pupils are to become familiar with the equipment, handle it with confidence, and succeed in getting the circuits functioning correctly. *The essence of teaching through assignments is pupil involvement.* Refer to Equipment Guide page 53 for details of equipment and the number of pupils that can be involved at any one time.

However, in *assignment 6*, pupils first analyse a complex circuit and after assembly check their conclusions. It would be very time-consuming for five groups of pupils to assemble this complex circuit so it is suggested that all pupils to assemble this complex circuit so it is suggested that all groups first study the circuit and that the assembly is carried out later as a class activity.

Since a certain amount of recording is involved, one set of equipment can be shared by two groups, one 'writing up' while the other is doing the practical work. The other groups can be involved with other assignments, as it not necessary, at this stage of the taneously. Therefore, while some pupils work through the *Pneumatic-control assignments*, others can be occupied on the *Resistance* or *Rectification assignments*. However, assignments should be tackled in the recommended order, e.g. the *Resistance* and *Rectification assignments* must be completed before the *Transistor assignments* are attempted.

Special Demonstrations

With groups working on different assignments it is impracticable to give special demonstrations to the whole class. Also, at this stage, pupils have greater experience of investigational work, and the assignments themselves include more detailed experiments which at the beginning of the course might have been teacher demonstrations. It seems unnecessary to include any special demonstrations in this section.

However, a visit to a local factory to see the use of pneumatic equipment, or the showing of a film on industrial applications might be considered.

Buffer Experiments

Groups will now be working at their own pace and on completion of one assignment will move on to the next. However, some buffer experiments are suggested which might be useful on completion of this section of the course.

Homework Questions

A selection of questions appears at the end of this section.

Overall Aim of the Assignments

To provide sufficient knowledge and experience in the use of pneumatic equipment to enable pupils to solve practical problems which might arise in project work. Pupils should be able to devise and assemble appropriate pneumatic circuits.

Pneumatic-Control Assignment 1

Aims

1   To familiarise the pupils with the on–off poppet valve, the various types of actuators available, and the symbol for each.

2   By a series of investigations to help pupils appreciate that the double-acting cylinder produces a useful force only if one side of the piston is exhausted while pressure is applied to the other. (To move the piston in the opposite direction the conditions must be reversed, and in this connection the term 'change-over' introduced.)

3   To introduce the principle of speed control.

Apparatus per group

one single-acting cylinder
one double-acting cylinder
two hand-operated on-off poppet valves
range of valves showing various types of actuators
one tee junction
five connecting pipes

Pneumatic-Control Assignment 2

Aims

To introduce four methods of controlling a double-acting cylinder.

1   By using a hand-operated, five-port double change-over valve.

2   By impulses from two hand-operated valves, in conjunction with a double pilot-actuated double change-over valve.

3   To provide automatic oscillation by positioning valves which are actuated by the piston rod itself.

4   To provide full control of the cylinder by inserting two extra valves to enable the piston rod to be stopped in either the retracted or the extended position, and to operate for one cycle only.

Apparatus per group

one double-acting cylinder
one hand-operated five-port double change-over spool valve
one double pilot five-port double change-over spool valve
one single pilot spring-return valve (for inspection only)
two lock-down on-off poppet valves
one roller-actuated on-off poppet valve
one plunger-actuated on-off poppet valve
one cross junction or two tee junctions
ten connecting pipes

Pneumatic-Control Assignment 3

Aims

To achieve control of piston speed of both double-acting and single-acting cylinders using exhaust restrictors and flow-control valves.

Apparatus per group

one double-acting cylinder ⎱ the larger the diameter, the
one single-acting cylinder ⎰ greater the degree of control
one double pilot five-port double change-over spool valve
one plunger-operated on-off poppet valve
one roller-operated on-off poppet valve
two exhaust restrictors
one flow-control valve
one cross junction or two tee junctions
eight connecting pipes

Pneumatic-Control Assignment 4

Aims

1   To introduce the three-port single change-over valve as an inverter (similar function to the relay used in *Electrical-switching assignment 3*).

2   To introduce the air reservoir and its use in pneumatic circuits.

Apparatus per group

one single-acting cylinder
one single pilot spring-return three-port spool valve
one reservoir
one lock-down on-off poppet valve
one diaphragm-actuated on-off poppet valve
one exhaust restrictor
one cross junction or two tee junctions
one tee junction
eight connecting pipes

Pneumatic-Control Assignment 5

Aims

To investigate further circuits involving the use of a reservoir and flow-control valves.

Apparatus per group

one single-acting cylinder
one flow-control valve
one reservoir
one single pilot spring-return three-port spool valve
one lock-down on-off poppet valve
two tee junctions
seven connecting pipes

Pneumatic-Control Assignment 6

In this assignment a complex circuit is presented for analysis. This will test the understanding of the previous work.

If sufficient equipment is available it is recommended that the pupils connect up this circuit in order to check their conclusions.

Experience has shown that a teacher demonstration of this circuit may be appropriate to assist pupils in their construction. Encourage pupils to construct in stages i.e to establish first the correct function of valves A, B, J, K and L and reservoir M and then to complete the remainder of the circuit.

Follow-up 6 suggests an alternative circuit — a circuit more likely to be used in industry. This employs sequence switching; the completion of one operation causes the next to commence — in this way the correct sequence of operations is always maintained.

Apparatus Required

one single-acting cylinder
one double-acting cylinder
one push-button on-off poppet valve
one plunger-actuated on-off poppet valve
one single pilot, spring-return three-port spool valve
two double pilot three-port spool valves

one double pilot five-port spool valve
three reservoirs
two flow-control valves
two exhaust restrictors
two cross junctions or four tee junctions
six tee junctions
twenty-four connecting pipes

For the circuit in *follow-up 6*, the following additional equipment is required.

one plunger-actuated on-off poppet valve
one one-way trip-actuated on-off poppet valve

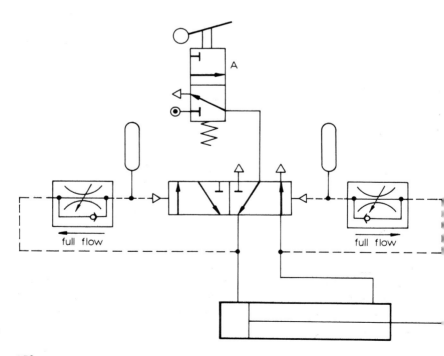

Homework Questions

1 Draw a diagram of a pneumatic circuit which will produce a delay between the time of operating a valve and the resulting movement of a single-acting pneumatic cylinder. How can the delay time be varied?
   If the maximum duration of this delay is to be increased, what modifications would you make to your circuit?

2 When a push-button valve is given an impulse, the piston rod of a double-acting cylinder is required to move out slowly and remain fully extended for three seconds before retracting quickly. Design a circuit to achieve this.

3 What will happen when valve A in the circuit shown on page 172 is switched on? Explain fully the operation of the circuit.

4 In the circuit shown below, which valves must be operated to make the piston rod extend?

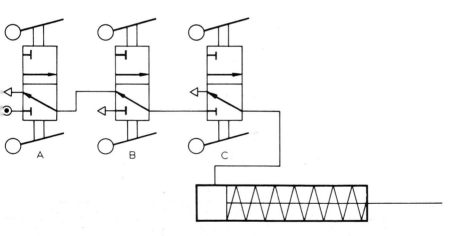

5 A door, operated by a double-acting pneumatic cylinder, is required to open when a person approaches it, and to remain in the fully open position for five seconds before closing. If other persons approach the door before it closes, it must remain fully open for five seconds after the last person has passed through.

   Design a pneumatic circuit to achieve this.

6 When an impulse is applied at valve A in the circuit below, the piston rod should immediately extend, and retract after a delay. What component is missing?

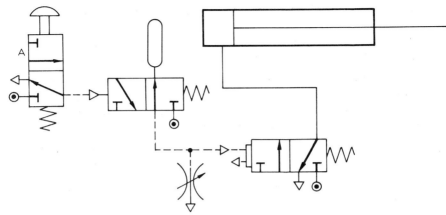

7 What will happen when valve A is switched on in the circuit shown below? Explain fully the operation of the circuit.

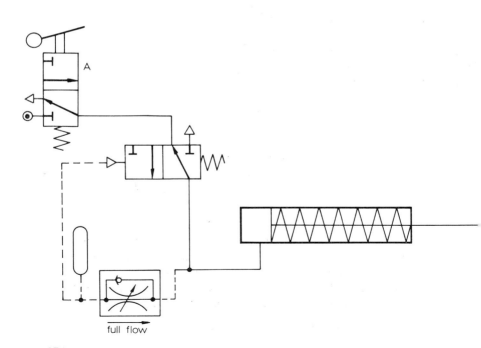

full flow

174

Buffer Experiments

1 The circuits given in Homework questions 3, 6 and 7 could be made up to verify the pupils' explanations.

2 The pupils' circuit designed in Homework questions numbers 1, 2 and 5 could be connected up and tested.

Electronics

General Introduction to Electronics Assignments

The purpose of this section of the course is to extend the pupils' knowledge of basic electrical and electronic circuitry in order that they may later devise suitable circuits for control purposes.

The *Electronics assignments* are divided into three sections: *Resistance Rectification* and *Transistors*. Originally this course made use of a specially designed turntable for assignment work. However many teachers feel that the turntable unit can be dispensed with for assignment work and replaced with an electric motor and a 360:1 ratio gearbox fitted with a Meccano wheel. Pupils' assignments are now written on this basis. Within the course the turntable unit serves two other important functions:

a) as the basis for mini or minor projects which a teacher may wish to set during the period — the turntable being a convenient unit involving, say, the removal of articles after being detected during sorting, counting, selecting, etc.;

b) to enable pupils to make preliminary investigations and to try out selected solutions to a major group project attempted at the end of the Transistor assignments — details of this project are given later.

NOTE: Buffer experiments are not included in these notes. At this stage in the course, the more capable pupils should be able to devise their own additional work.

Electronics: Resistance Assignment 1

This assignment is concerned primarily with a simple method of reducing the speed of a d.c. motor by using resistors of various values in series with the motor. Pupils discover that the effective resistance of two resistors placed in series is the sum of individual resistors.

The potential-divider principle emerges from the results of the various investigations, and the follow-up indicates how a variable resistor can be used to give a variable output voltage from a fixed-voltage supply. The limitations are not discussed, though teachers should be aware of them. The two main disadvantages are explained below with the help of diagrams

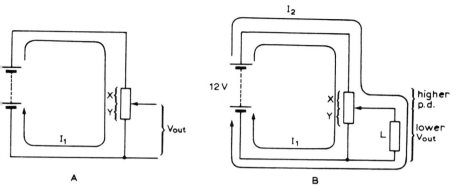

A   **B**

i) Diagram A shows that, whenever a potential-divider is set up, a current I_1 is taken from the supply, irrespective of whether or not current is being delivered to an external circuit.

ii) Diagram B shows the effect when an external circuit is added. A second current path is produced, I_2 representing the current taken by the device L. If the slider is central, resulting in equal resistances above and below it, then with no load (diagram A) the voltage appearing across both X and Y will be 6 V, since currents in X and Y are identical.

p.d. across X $= I_1 \times R$

p.d. across Y $= I_1 \times R$

When an external circuit is completed, however, as when supplying an electric motor from the divider, an additional current flows in X but not in Y. Thus the p.d. across X must now be greater than that across Y, say 8 V and 4 V respectively, and the output voltage falls. The output voltage of a potential-divider network is, therefore, very sensitive to the current taken by the device being supplied.

Apparatus Required (15 pupils, 5 groups of 3)

five electric motors fitted with a 360:1 ratio gear box
five Meccano road wheels
five power supplies — 12 V d.c. 2 A — with safety fuse or trip

ten multimeters, or five 0–20 V volmeters and five 0–1 A ammeters
ten 25 ohm resistor units
five 50 ohm resistor units
stackable 4 mm plug leads — assorted lengths

Demonstrations

1   Show pupils a variety of wire-wound resistors — fixed and
    variable. Discuss the reasons for the differences in size.

2   If pupils are not familiar with the use of a multimeter, then
    demonstrate its use now, leaving ohmmeters until *assignment 2*.

Homework Questions

1   Explain why the speed of a motor is reduced when a resistor is
    placed in series with it.

2   A resistor of 100 ohms is placed in series with a 12 volt motor.
    If, while running, the motor has an effective resistance of
    200 ohms, what voltage appears across the motor?

3   Two 12 volt motors are fitted in series across a 12 volt d.c.
    supply.

    a) How will the speed of each motor compare with speed of
       such a motor placed directly across a 12 volt d.c. supply?

    b) If a voltmeter is fitted across each of the motors placed in
       series, what is each meter likely to read?

    c) What is the effect if the two motors in series are placed
       across a 24 volt d.c. supply?

Electronics: Resistance Assignment 2

Pupils are asked to compare the resistance of a light-source
filament when cold with its resistance when hot. They later use a
variable resistor to control the speed of an electric motor.

The ohmmmeter is introduced as a direct means of resistance
measurement, and suitable meters should be available. Teachers
should demonstrate to the class how these meters are used.

The follow-up draws attention to the limitations of variable-
resistance motor control, namely that the high starting current

equired by the motor results in a low voltage supplied to it at this
ime.

Apparatus Required (15 pupils, 5 groups of 3)
ive electric motors fitted with a 360:1 ratio gearbox
ive Meccano road wheels
ive power supplies — 12 V d.c., 2 A — with safety fuse or trip
ive 0–2 V voltmeters
ive 0–1A ammeters
ive lamp indicator units
ive 100 ohm variable-resistor units (or similar)
ive ohmmeters giving an appreciable reading for 100 ohm (one
vould be sufficient if passed round to each group)
tackable 4 mm plug leads — assorted lengths

Demonstrations

Demonstrate the use and discuss the construction of an
ohmmeter to support the assignment notes.

Demonstrate the use of a variable resistor, connected as a
potentiometer, on an a.c. supply. Show that the a.c. output
voltage — as measured on an a.c. voltmeter — varies with the
position of the slider in a similar way to d.c.

Demonstrate the use of a potentiometer as a volume control in
a radio receiver. In this case the a.c. signals from the receiver
appear across the whole control, the output being taken
between one end and the slider.

Homework Questions

How does that resistance of the filament of a tungsten lamp
vary with its temperature? Such lamps eventually 'blow' when
they are first switched on. Why is this?

When an ammeter is placed in series with an electric motor and
a power supply, the motor speed falls slightly.

a) Give an explanation for this.

b) Is the reading on the ammeter a reliable indication of the
current taken by the motor when first connected to a 12 volt
d.c. supply? Provide an explanation.

Imagine you have been given a milliammeter with a blank scale.

Describe how you would make an ohmmeter from the instrument, and describe one way of roughly calibrating it, assuming you had a supply of accurate resistors.

4   You are given a 12 volt d.c. power supply, variable resistor, and a 12 volt electric motor. Draw circuit diagrams to show how you would connect the variable resistor, firstly as a rheostat and secondly as a potentiometer. One of the arrangements is less likely to enable you to stop the motor or to run it at full speed. Which one is this, and why?

Electronics: Resistance Assignment 3

In this assignment, pupils compare the output voltage of a mains-operated low-voltage power supply with that of a lead–acid accumulator under different loading conditions. They should notice that the output voltage falls with increasing current demand; this is evident when using a mains unit even if relatively small currents are delivered to a load. This 'loss' of voltage from a power supply may suggest to pupils that there must be a resistance 'built into the supply' which forms a potential-divider with the external load. The follow-up explains the effect in detail.

This assignment also introduce the principle of 'power rating'. It is essential that pupils should be aware of heat generation in electronic apparatus whenever electrical energy is converted into heat energy.

Note that some teachers have reported problems with this assignment saying that the applied load produced no significant effect. It is important to remember that the applied load must be significant in terms of the maximum rating of the supply. If heavy duty supply units are used, for example wall mounted laboratory suplies or 10–20 A bench supplies, then the effect of the suggested loads will be very small. A second reason for the lack of results may be the use of stabilized or regulated supplies. These are designed to compensate for the effects of the applied load and will normally show a voltage drop of less than 1% even at full load.

If you do not have a suitable set of supplies, they can easily be made by fitting a rectifier to the transformer units used in the rectification assignments. (See Equipment Guide for details).

Apparatus Required (15 pupils, 5 groups of 3)

Five power supplies — 12 V d.c. 2 A — with safety fuse or trip
0–2 V voltmeters
Five high-current solenoids (one or two may be sufficient)
One fully charged lead-acid accumulator, 12 V car-battery capacity
Ten 25 ohm resistor units (5 W) } 25 ohm, 50 ohm, 75 ohm, and
Five 50 ohm resistor units (5 W) } 100 ohm can be obtained
from these
Five 10 ohm, approximately ¼ W carbon resistors (expendable)
Five 10 000 ohm, ¼ W carbon resistors
Takable 4 mm plug leads — assorted lengths

Demonstration

Discuss the precautions which must be taken when using photocells
in series with a low-resistance device. For instance, one of the
electrical-switching assignments suggests fitting a photocell and
lamp indicator unit in series with a 12 volt supply. Under this
arrangement the cell is working at almost maximum heat
dissipation. A higher current lamp could cause the cell to be
damaged.

Homework Questions

The circuit over page shows a photocell fitted in series with a
lamp and a 12 volt supply. When the cell is illuminated, the
resistance of the cell is 60 ohm and that of the lamp 120 ohm.

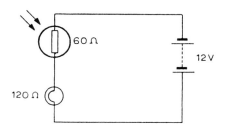

a) How much current flows in the cell and in the lamp?
b) What voltage appears across the cell and across the lamp?
c) If the cell is rated at 0.25 watt (250 mW), is it likely to be
   damaged?

2   An electric fire is connected to a 250 volt mains and carries a current of 4 amperes. What is the wattage rating of the fire?

3   A 3 kW elecric kettle is connected to a 250 volt supply. What current is drawn?

4   The electric motors you use sometimes become warm. Why is this, and under what kind of conditions is it likely to overheat?

Resistance Assignment 4

The resistor colour code is introduced as a means of identifying values of resistors used in electronic equipment, the code being mainly used for the low-wattage types. It is not envisaged that pupils will memorise the code, but it should be made available for easy reference.

Pupils try connecting resistors in parallel, to determine the effective value of the arrangement. The relationship is not easily deduced, and a method of calculating the effective resistance of a parallel (shunt) arrangement is given in the follow-up.

Apparatus Required (15 pupils, 5 groups of 3)

fifty colour-coded carbon resistors of assorted values and tolerances. If preferred, use five sets of ten assorted types in order that each group has the same values, though this is not essential. Use a large spread values from less than 10 ohm to a few megohm; tolerance values should include 5% (gold band), 10% (silver band) 20% (no band), and if possible 1% (brown band) and 2% (red band).
five 50 resistor units (5 W)
five ohmmeters giving large readings in the region of 100 ohm
stackable 4 mm plug leads — assorted lengths

Demonstrations

1   With more able groups, show that the colour coding of capacitors is similar to that for resistors.

2   Discuss the components in a chassis containing electronic components. Identify resistors, and discuss their values and likely wattage ratings.

Homework Questions

1 You have a supply of 10, 20, and 50 ohm resistors. Using these values only, describe the arrangements you would use to obtain resistances of:

   a) 30 ohm,

   b) 5 ohm,

   c) 125 ohm.

2 a) The coil in a relay unit has a resistance of 185 ohm. What current does the relay coil carry when it is connected across a 12 volt supply?

   b) Explain as fully as you can why any two relays may have different resistances.

3 A colour-coded resistor has red for the first digit, violet for the second, and orange for the multiplier.

   a) What is the nominal value of the resistor?

   b) If the tolerance band is coloured gold, between what two values of resistance would the actual value fall?

4 A 100 ohm, ¼ W resistor is placed directly across a 12 volt supply. Is the resistor likely to overheat?

Electronics: Rectification

Students will have been using low-voltage, mains-operated power-supply units during the course, but little has been said so far concerning the construction of such equipment. This section is intended to enable pupils not only to understand power-supply circuitry but also to enable them to construct a power-supply unit, should the need arise. While it is not recommended that pupils should construct mains power supplies, owing to the danger involved with high voltages, it is possible that, during project work, d.c. will be required from a low-voltage a.c. source.

Introduction — The Use of the Cathode-Ray Oscilloscope

Little needs to be said concerning this combined assignment and follow-up except to stress that oscillospopes capable of accepting d.c. inputs are essential. In the assignments that follow, oscilloscopes capable of displaying only a.c. are suitable, but, in order to give students an understanding of the principle of the oscilloscope, direct-current voltages are investigated in this first instance. If only one d.c. instrument is available, then this could be demonstrated.

Apparatus Required (15 pupils, 5 groups of 3)

one cathode-ray tube from an oscilloscope, or an oscilloscope with the cover removed in order that the tube can be seen (the latter is less satisfactory, since the tube will be shrouded in a mu-metal screen.)

five oscilloscopes, with at least one having facilities for d.c. inputs; pupils must use the d.c. input type for *investigation 5*.

five batteries to give 'pure' d.c., preferably 6 to 12 volts, though 1.5 volt torch cells could be used.

stackable 4 mm plug leads — assorted lengths. Ensure that pupils can make good electrical connections to the oscilloscope inputs if these are not 4 mm size.

Demonstrations

1   If possible, show the pupils a cathode-ray tube of the type used in an oscilloscope (electrostatic-deflection type), pointing out the relevant parts. If this is not possible, remove the cover from an oscilloscope in order to see the tube.

2   If only one d.c. input oscilloscope is available, teachers may wish to demonstrate 5 and 6, though it is more satisfactory to allow the pupils to attempt these sections themselves.

Homework Questions

1   Assuming that a focused spot is required at the centre of an oscilloscope screen, which controls would you adjust if you observed the following:

a) the spot is central on the screen but is blurred?

b) the spot is sharp but positioned to the left of centre (at nine o'clock)?

c) the spot is sharp but positioned above centre (at twelve o'clock)?

d) the sport is sharp but dim and positioned at two o'clock?

2 An oscilloscope is set up so that there is a horizontal line on the face of the tube. If it takes 0.01 seconds for a spot to trace out this line,

a) how long it take to trace out half of the line?

b) how many line are traced out each second?

3 An oscilloscope is set up with a line on its screen. Draw a diagram to indicate the change of position of the trace if,

a) a 3 volt battery is fitted across the Y input terminals, with the oscilloscope adjusted for d.c. inputs;

b) the 3 volts is increased to 6 volts, then to 12 volts.

4 A spot is placed centrally on the screen of an oscilloscope. The circuit below is set up, and the output is connected to an oscilloscope as shown. What would you expect to see on the screen when switch S is closed?

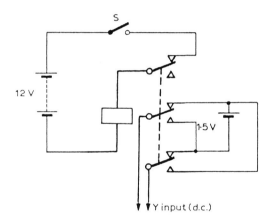

Electronics: Rectification Assignment 1

Assignment 1 introduces the transformer and the diode, and the unidirectional property of the latter investigated by the pupils. For the first time, pupils briefly connect a d.c. motor to an a.c. supply and note the effect. The resulting vibration, rather than continuous motion in one direction, should lead pupils to conclude that the motor is attempting to rotate first in one direction and then in the other. This conclusion is confirmed when the diode is placed in

185

series with the a.c. supply. This device first appears to block that portion of the alternating supply which produces current flow in one direction, and then, after reversal, that part of the alternating supply which causes current to flow in the other direction. Teachers should ascertain that all pupils have a clear understanding of alternating current at the completion of this assignment.

Apparatus Required (15 pupils, 5 groups of 3)

five transformer units (mains input, 12–0–12 V output)
five electric motors fitted with a 360:1 ratio gearbox
five Meccano road wheels
five power supplies — 12 V d.c., 2 A — with safety fuse or trip
five oscilloscopes
five 50 ohm resistor units (5 W)
five diode units
five 0–20 V voltmeters
stackable 4 mm plug leads — assorted lengths

Demonstration

Exhibit different types of diode when pupils have completed the assignment. Show low-current as well as high-current types. If possible, show the older type of 'finned' metal diode (rectifier). Discuss the relative sizes, e.g. semiconductor diodes of the type used in the diode unit are considerably smaller than the 'finned' selenium type. If possible, also compare the physical size of valve rectifiers with semiconductor diodes of the same current-carrying capacity.

Homework Questions

1 A 12 V 0.1 A bulb is fitted into a lamp indicator unit and the latter is then placed across the 12 volt out put of a transformer connected to the 240 volt mains. what current flows in the primary coil of the transformer?

2 The primary coil of a transformer has 2000 turns, and the secondary 50 turns. If the primary coil is connected to the 240 volt mains, what voltage is obtained from the secondary?

3 A diode is suspected of being faulty. Describe how you would use an ohmmeter to test it. What kinds of readings would you expect on the meter if:

a) the diode is 'good'?
b) the diode is faulty?

In the circuit shown below, what are the possible causes if:

a) the motor
   'buzzes' but
   does not
   rotate?
b) the motor
   does not
   'buzz' or
   rotate?

240 V a.c.   12 V a.c.   (M) motor

Electronics: Rectification Assignment 2

One cannot expect pupils to 'discover' or design a full-wave rectification circuit, and in this assignment, therefore, circuits for full-wave and full-wave 'bridge' rectification are given. Comparisons are first made between the waveforms obtained from half-wave, full-wave, and full-wave 'bridge' circuits. Each supply in turn is connected to the motor, and it will be noted that the two full-wave circuits result in very similar motor speeds; but the motor runs noticeably slower when using the half-wave circuit. Reference to the waveforms should enable pupils to offer a possible reason. Careful questioning should produce the explanation that each peak of the wave form represents a voltage 'pulse' which gives the motor a 'push', or, better still, each pulse supplies energy to the motor. Since there are twice as many pulses per second using full-wave (100) than when using half-wave (50) rectification, the motor speed changes — i.e. the average voltage supplied from a full-wave circuit is higher than the average voltage from a half-wave circuit.

In order to assist pupils in understanding the function of the extra diodes used in full-wave circuits, they should observe the waveforms indicated across each half of the secondary of the transformer as used in the two-diode circuit. The success of this test will depend upon the type of oscilloscope available. Some success is achieved if synchronisation is removed from the oscilloscope and the trace made as near stationary as possible by means of the fine velocity control. Each half of the transformer is then tested in turn.

To obtain a better result arrange a switching circuit, using double-pole double-throw unit, to connect each end of the transformer in turn to the oscilloscope Y input. The centre of the winding of the secondary is connected to the 'earth' (common) terminal. By rapid switching, the difference between the two traces is evident.

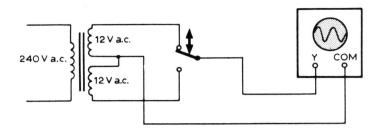

When correctly set up, the waveforms should appear 180° out of phase:

1st observation   2nd observation

A still better result is achieved using a double-beam oscilloscope, enabling the two traces to be viewed simultaneously.

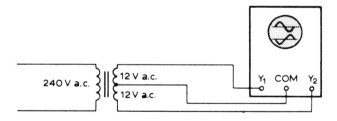

In 7, the motor is loaded by simply applying a finger as a 'brake' to the rim of the wheel. Pupils should note the increase in the reading of current to the motor — an important fact in connection with the next assignment.

Apparatus Required (15 pupils, 5 groups of 3)

five transformer units
twenty diode units

188

five 50 ohm resistor units (5 W)
five oscilloscopes
five Meccano road wheels
five electric motors fitted with a 360:1 ratio gearbox
five 0–1 A ammeters

electric motors fitted with a 360:1 ratio gear box

Note *investigation 6* is best performed using a double-beam oscilloscope, in order to observe the two waveforms simultaneously.

Demonstration

Demonstrate the use of the double-beam oscilloscope in order that the waveforms at each end o the transformer secondary can be seen simultaneously.

Homework Questions

1   Illustrate two major differences between full-wave 'bridge' rectification and full-wave rectification.

   Modern miniaturised electronic equipment usually employs the 'bridge' rectification circuit. Why do you think this is?

2   Design a battery charger circuit, with the following features:

   a) a means of measuring the charging current,
   b) a control to vary the charging current,
   c) a pilot lamp to indicate that the unit is switched on,
   d) an on-off switch,
   e) a means of cutting off the supply from the battery if the battery being charged suddenly develops a short-circuit.

3   Design a power supply to give outputs of 6 volts d.c. and 24 volts d.c. from the same transformer.

4   One of the diodes becomes faulty in a full-wave rectifier circuit. What will the effect be if:

   a) the diode goes short-circuit?
   b) the diode goes open-circuit?

Electronics: Rectification Assignment 3

For most applications, mains-derived d.c. power supplies should be 'smooth'. This is particularly important when the supply is used to power an electronic circuit. A smooth supply, approximating to the d.c. provided by primary cells (e.g. torch cells) or secondary (rechargeable) cells (e.g. motor-car accumulator) is often essential. The d.c. obtained by pupils in the previous assignments is far from smooth, being in the form of pulses.

Graphical representations of 'smooth' and 'pulsed' d.c.

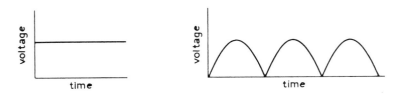

The purpose of *assignment 3* is to attempt to produce smooth d.c. supplies by fitting 'smoothing capacitors', and to discover the factors which determine the value of the capacitor required. Pupils will discover that, for a given value of smoothing capacitor, the larger the current taken from the supply, the greater is the ripple. The optimum value selected for a particular purpose is determined by the current consumed and the allowable ripple content of the d.c. output.

Apparatus Required (15 pupils, 5 groups of 3)

twenty diode units
five transformer units
five 5000 μF capacitors (or any value in this region)
five 1000 μ capacitors (or values in this region)
various capacitors, ranging from 100 μF to 10 000 μF
five oscilloscopes
five 50 ohm resistor units (5 W) } these, in series and parallel,
ten 25 ohm resistor units (5 W) } will provide a range of values.
stackable 4 mm plug leads — assorted lengths
various capacitors, ranging from 100 μF to 10 000 μF

Demonstrations

1 If possible, show the layout and components of a power supply incorporated in a piece of electronic equipment.

2 By group discussion and demonstration, summarise the *Rectification assignments.*

Homework Questions

1 Draw sketch graphs to represent the output from an unsmoothed rectified supply and from a battery used in a torch. Explain the differences between the two graphs.

2 The value of the output voltage from an unsmoothed rectification circuit, as measured on a voltmeter, is different from that of a similar circuit incorporating a smoothing capacitor. Why do these voltages differ?

3 Two separate power supplies, each consisting of a transformer and a full-wave rectifier, supply 3 amperes at 12 volts to external circuitry. If a 'smooth' output is essential, which will require the smallest smoothing capacitor, assuming one operates from 250 volts, 50 Hz and the other from 250 volts, 500 Hz?

4 'The greater the current taken from a smoothed rectification circuit, the greater is the ripple content of the output.' Give an explanation for this effect.

Electronics: The Transistor

Pupils have already become acquainted with one semiconductor device, the semiconductor diode, and they are now introduced to a second, more versatile, one: the transistor. As a result of this series of assignments, pupils become familiar with some of the more important properties of the transistor. *Assignment 4* asks pupils to modify an existing circuit to produce a light-operated switch incorporating a transistor. Having successfully completed the investigation, pupils have a practical circuit suitable for use in later project work. It is anticipated that, when all five assignments have been completed, teachers will introduce pupils to more complex circuitry, and perhaps discuss alternative methods of assembly.

For a number of these assignments the pupils use a 25 kilohm potentiometer, either as a rheostat or as a potentiometer. The value of this component is not critical. Many schools will already have 50 kilohm units. These can be used for all the assignments. The 25 kilohm units now recommended do give slightly better results and it is suggested that unserviceable 50 kilohm units are replaced by these units.

Electronics: The Transistor Assignment 1

Here pupils examine a transistor unit containing a power transistor, and names are given to the three connections. The teacher should have several different types of transistor available for the pupils to examine. It should be emphasised that the transistor itself is usually much smaller than its package suggests. It is possible to remove the top from a type contained in a metal can in order to make the construction visible. A binocular microscope may be necessary. Some transistors encapsulated in a metal case are coated in silicon grease, which is rather difficult to remove completely.

Note that beryllium oxide is a good electrical insulator but has a thermal conductivity equal to that of pure copper. For this reason beryllium oxide is used in some modern high-powered semiconductor devices. Beryllium oxide is amongst the most toxic materials known. In the long term people's health can be severely damaged by only a few micrograms of the dust in the lungs.

While this material is not often used, and cutting open transistors up to the size of 2N3055 is safe, great care should be taken to ensure that pupils never cut open unknown devices. For instance a 78 series voltage regulator looks (apart form the marking) identical to some 2N3055 transistors.

The three parts of the transistor — base, collector, and emitter — are easily identified in 'alloy' constructions.

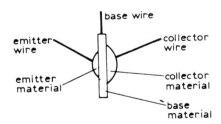

The diagram shows the construction. Note the size of the collector region compared with the emitter. It is recommended that teachers should refer to suitable literature on transistor construction and manufacture for background information.

Using the power transistor, pupils investigate the following.

1   The resistances measured between the base and emitter and between the collector and base. The measurements are repeated with the ohmmeter leads reversed. From their previous knowledge of diode action, pupils will note that a transistor in some respects resembles two diodes 'back-to-back'. This information provides a useful means of testing a transistor which is suspected of having been damaged by misuse. (It is assumed that a transistor tester is not available.)

2   The effect of omitting a base connection when a lamp is fitted in the collector circuit. The lamp will normally remain unilluminated, suggesting that either no current is passing between collector and emitter or else a current which is insufficient to light the lamp.

3   The effect on the emitter-collector current when a base current flows. In *investigation 6*, a base current is provided by means of a 1.5 volt cell. Pupils will note that the lamp in the collector circuit is illuminated, whereas the lamp in the base circuit remains unlit. This suggests that the base current is considerably less than the collector current. This is confirmed when ammeters are fitted in the base and collector circuits. Reversing the 1.5 volt cell prevents the lighting of the lamp in the collector circuit. This emphasises the importance of connecting the base supply voltage such that the base is positive with respect to the emitter when using the npn-type transistor.

Apparatus Required (15 pupils, 5 groups of 3)

five power-transistor units
assorted transistors for demonstration, including other power and small signal types
five ohmmeters
five diode units
ten lamp indicator units
five 1.5 V cells
five power supplies — 12 V d.c., 2 A — with safety fuse or trip
five manual switch units
five multimeters or five 0–0.5 A ammeters
stackable 4 mm plug leads — assorted lengths

Demonstration

Show a number of transistors, including both high-power and low-power types illustrating that pnp and npn types may look identical. Discuss how to use an ohmmeter in order to determine whether a transistor is npn or pnp. Refer to the diagram in *follow-up 1, part 3*, during the discussion.

Homework Questions

1   High-power diodes and transistors are often mounted on 'heat sinks'. What is the purpose of a heat sink, and what is the best material to use for the purpose?

2   You are give two unmarked transistors; one is known to be a npn type and the other pnp. What test would you use to discover which is which?

3   In *Transistor follow-up 1* it is suggested that a power transistor can be likened to a relay, in that both can act as switches. Suggest one advantage in using a transistor rather than a relay. Give reasons.

4   In circuit diagrams, how are npn transistors distinguished from pnp types?

Electronics: The Transistor Assignment 2

A variable resistor is fitted into the base circuit of the transistor, in addition to the 1.5 volt cell. A lamp fitted in the collector-emitter circuit varies its brightness as the variable resistor is adjusted — diagram (a). In order to dispense with one of the supplies (the 1.5 volt cell), the base current is then provided from the same supply as that feeding the collector — diagram (b) below.

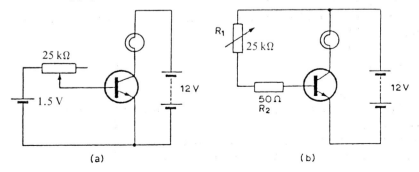

(a)   (b)

It must be stressed that if the fixed resistor R_2 is omitted, then damage to the transistor may result. Should the variable resistor be set to zero, in the absence of R_2, the 12 volt supply will be connected directly across the base and emitter of the transistor. This situation may damage the base–emitter region or, alternatively, a large collector–emitter current may be produced which will cause the transistor to overheat. Should the power transistors become damaged despite warnings, the following modifications to the transistor unit are recommended:

a) Fit a 50 ohm, 5 watt wire-wound resistor between the green socket and the base of the transistor. This will afford considerable protection.

b) Fit a 5 ohm, 5 watt wire-wound resistor between the red socket and the transistor collector. (Ideally this resistor should be of about 30 watt rating to meet all conditions, but such a resistor would be too large to be fitted in the unit.) This resistor will afford some protection, but it or the transistor may still overheat if large collector currents are allowed over long periods of time.

Pupils replace the variable resistor by a photocell and are asked to use the transistor first as a switch and then as a variable resistor. In the first instance, either full or no photocell illumination is required. This will result in the transistor passing either a high current or almost none. In the second case, the transistor acts rather like a variable resistor — in the sense that the current between collector and emitter can be varied. The transistor collector–emitter variations can be produced by varying the intensity of the illumination of the photocell by a suitable means.

Apparatus Required (15 pupils, 5 groups of 3)

five power supplies — 12 V d.c., 2 A — with safety fuse or trip
five power-transistor units
five 25 kilohm variable-resistor units
five 1.5 volt cells
five 50 ohm resistor units (5 W)
five lamp indicator units
five 0–1 A ammeters
five electric motors fitted with a 360:1 ratio gearbox
five 50 ohm variable-resistor units — or 100 ohm (high wattage)
five photocell units

five light-source units
stackable 4 mm plug leds — assorted lengths

Demonstration

Remove the outer casing from a small transistor in order that pupils
can identify the base, emitter, and collector regions under a
microscope (preferably binocular).

Homework Questions

1 What is the purpose of the 50 ohm resistor in the circuit shown
   below?

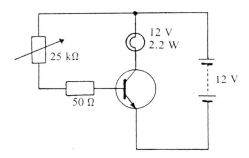

2 With reference to the circuit in question 1, if the lamp glows
   normally, what current is flowing in the collector circuit of the
   transistor?

3 A base current of 20 microamperes is found to produce a
   collector current of 1 milliampere. What is the current gain
   (H_{FE}) of the transistor at this collector current?

4 A 0.1 A lamp fitted in the collector circuit of a transistor is to
   be illuminated fully when the transistor is conducting. If the
   current gain of the transistor is 100, what current must flow in
   the base circuit?

Electronics: The Transistor Assignment 3

The purpose of this assignment is to make pupils aware of two
basic characteristics of a transistor

a) Current gain (h_{FE}). The calculation of this is given in the follow-
   up.

b) The base–emitter current for a silicon transistor is extremely low until the base voltage exceeds about 0.6 V. Once the base has reached its 'turn-on' voltage its resistance becomes very low and a small increase in voltage produces a large increase in current.

The follow-up material includes examples with a simple potential divider to illustrate two factors which must be taken into account when designing circuits,

a) The input voltage must exceed 0.6 volts.

b) the input must be able to supply sufficient current to the base.

Apparatus Required (15 pupils, 5 groups of 3)

five 0–1 A ammeters
five 0–10 mA meters ⎫ or similar ranges
five 0–1 mA meters ⎭
five power supplies — 12 V d.c., 2 A — with safety fuse or trip
five power-transistor units
five 25 kilohm variable-resistor units
five 50 ohm resistor units (5 W)
five 1.5 volt cells
five relay units
stackable 4 mm plug leads — assorted lengths

In this assignment pupils are asked to set certain potentiometer positions. You may find that it helps them if you paint index lines on the knobs of the 25 kΩ potentiometers and fix a scale, similar to the one below, to each unit.

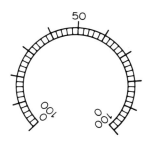

Demonstration

Set up the transistor tester shown below in order to test the current gain of an assortment of transistors. It will also determine whether or not a transistor is faulty.

In order to test pnp transistors, the batteries and meters must be reversed. It is recommended that the circuit is wired up permanently in a suitable case, enabling pupils to test a transistor quickly, without the necessity of setting up the arrangement each time. In a permanent unit, an on–off switch should isolate one lead from each

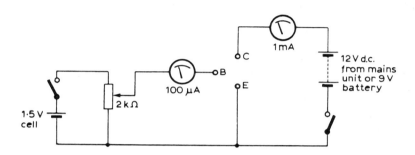

battery. Three terminal posts should be provided to make good electrical contact with the transistor leads, and ideally a change-over switch be incorporated to reverse the meter and battery connections for testing both npn and pnp transistors.

Homework Questions

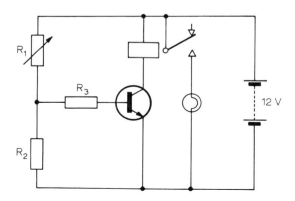

1 The circuit shown above is set up and it is found that the relay does not change over to the energised condition in any position of the variable resistor. Give *four* possible component failures (other than wiring) which could produce this result. Suggest a reason in each case.

2 A wire-wound variable resistor is suspected of being faulty. How would you test it in order to pronounce it 'serviceable' or 'unserviceable'?

3 Describe the behaviour of the lamp in the circuit below when the switch S is closed. Give reasons for your answers.

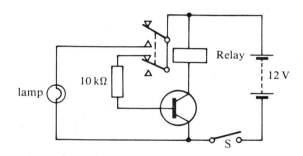

Electronics: The Transistor Assignment 4

When pupils use a photocell in the base circuit of the transistor, a very sensitive light-operated switch is produced, and several investigations are suggested. By reversing the positions of the photocell and the fixed resistor in the potential-divider feeding the base circuit of the transistor, the operation of the circuit becomes reversed. In the first case — the photocell connected between supply positive and base — the relay is energised when the cell is illuminated and de-energised when the light incident upon the cell is blocked. On reversing the two components the relay is de-energised when light is incident upon the cell and energised when the light is blocked.

Note that in this assignment it is important to draw the pupil's attention to the 1 kilohm resistor in series with the photocell. This protects it from the large currents which could flow if a light is shone on the photocell while the rheostat is turned to a very low value.

Having completed this series of assignments on the transistor, and having been introduced to useful light-operated and temperature-operated switches, investigations involving more advanced electronic circuits should be feasible with the more-able pupils.

Apparatus Required (15 pupils, 5 groups of 3)

five lamp indicator units
five 50 ohm resistor units
five 25 kilohm variable resistor units
five power supplies — 12 V, 2 A — with safety fuse or trip
five photocell units
five light source units
five two-pole two-way relay units
five electric motors with a 360:1 ratio gearbox
five power transistor units
stackable 4 mm plug leads — assorted lengths

Demonstration

During project work students are often asked to produce time delays. They will already have met two types of relay delay circuits — one in Electrical switching Assignment 5 and another suggested in the teachers' notes for that assignment. The latter uses a variable resistor in series with a relay coil shunted by a capacitor. If this circuit has not been demonstrated, teachers may wish to do so now. An alternative transistor circuit is shown below. Since it may enable pupils to be more versatile when dealing with design problems, it is recommended that the teacher should demonstrate the technique to pupils of suitable ability.

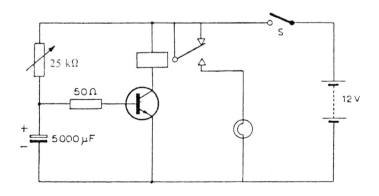

When switch S is closed, the current flowing in the variable resistor begins to charge up the capacitor in preference to causing a base current to flow (an uncharged capacitor behaves as a short-circuit). A base current begins to flow when the capacitor has charged sufficiently, and hence a collector current flows. Eventually the collector current is high enough to energise the relay, and the lamp illuminates. Increasing the resistance of the rheostat increases the delay time by decreasing the charging rate of the capacitor.

In the follow-up material to this assignment the pupils are introduced to the 'Darlington Pair' transistors which can be used as a 'super-gain' transistor. It may be worthwhile to make up a few 'Danum Trent' modules with Darlington Pair transistors. These can easily be made and fitted into a transistor unit.

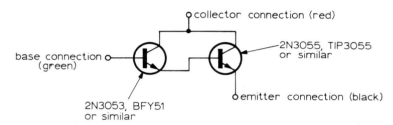

collector connection (red)

base connection (green)

2N3055, TIP3055 or similar

2N3053, BFY51 or similar

emitter connection (black)

Homework Questions

1   If you were to set up the following circuit, what effects would you expect if the light from one or both the light sources were to be blocked?

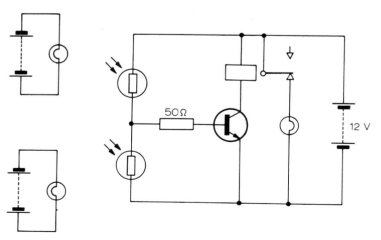

50 Ω

12 V

2   When light strikes the photocell in the circuit shown below, what approximate voltage reading would you expect on the voltmeter? If the light were to be prevented from reaching the cell, what would you expect to happen to the voltmeter reading? (*Hint*: What is the resistance between the collector and emitter of a fully conducting transistor compared with a non-conducting transistor?)

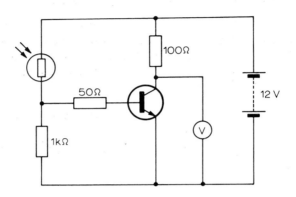

3   In the following circuit, the diode in the supply line acts as a safety device. For what specific purpose is this diode likely to be included?

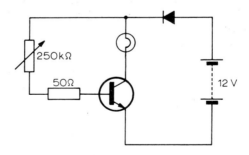

4   Redraw the following circuit for use with a pnp transistor.

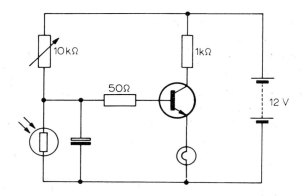

Electronics: The Transistor Assignment 5

In this last assignment the pupils are introduced to a new device, the thermistor, and are asked to design a circuit which uses it.

The pupils first set up a potential divider with the thermistor unit and a 25 kilohm rheostat. They investigate the effect of heating the thermistor and should appreciate that its resistance falls as it is heated and should note the similarities to a light-dependent resistor.

The pupils are then asked to design a high-temperature alarm. For this they may choose to use a single transistor but could be encouraged to consider using a Darlington Pair.

Apparatus Required (15 pupils, 5 groups of 3)

five lamp indicator units
five 50 ohm resistor units
five 25 kilohm variable resistor units
five 1 kilohm resistor units
five power supplies — 12 V, 2 A — with safety fuse or trip
five thermistor units (KR 472 CW)
five two-pole two-way relay units
five Darlington Pair power transistor units (if available)
stackable 4 mm plug leads — assorted lengths

Logic

In logic circuits pupils are made aware of the basic principles of logic by the use of mechanical switches, integrated circuits and logic gates built with transistors, diodes and resistors from the Danum Trent kit. *Assignment 3* and *5* use integrated circuit devices.

The assignments are written for use with any logic chips. However experience has shown that 7400 series TTL logic devices produce the most reliable results and we recommended that these be used. A problem arises because these devices require a 5 volt stabilised supply. The Equipment Guide gives details of suitable logic units and power supply module.

In these assignments the operation of the logic circuits is described with equation using Boolean notation. The symbols are., which means AND, + , which means OR, and a bar over the letter, which means NOT. For example A.B̄ means A AND NOT B.

Logic assignment 1

In *assignment 1* pupils construct simple logic gates with mechanical switches and produce truth tables for the arrangements they build. By the end of this assignment they should be aware of the basic functions of AND, OR, NAND, and NOR gates.

These concepts are reinforced in later assignments where they again meet these gates in integrated circuit form and build up the gates with discrete components. In *assignment 5* pupils are expected to apply these ideas to the solution of set problems.

At this stage you could suggest to pupils that pneumatic valves can also be used to produce logic gates. The four basic gates can be built up from three-port valves and shuttle valves as shown below.

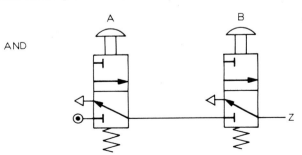

AND

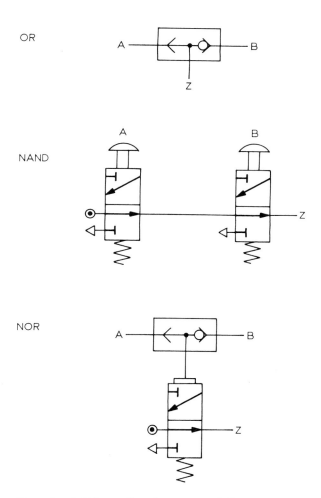

OR

NAND

NOR

Apparatus Required (15 pupils, 5 groups of 3)

ten SPST Push switches (DPDT Toggle switches could be used)
five lamp indicator units
five power supplies — 12 V, 2 A with safety fuse or trip
stackable 4 mm leads — assorted lengths.

Logic Assignment 2

In this assignment, pupils extend the work of *assignment 1* to look
at the logical equivalence gate and the exclusive OR gate.

Apparatus Required (15 pupils, 5 groups of 3)

ten SPST Push switches (DPDT Toggle switches could be used)
five lamp indicator units
five power supplies — 12 V, 2 A with safety fuse or trip
stackable 4 mm leads — assorted lengths.

Logic Assignment 3

Logic circuits 3 introduces pupils to logic gates using integrated
circuits. They again consider the four basic gates and the inverter or
NOT gate.

At this point strongly emphasise the need for a 5 volt supply for the
logic units. Then allow them to work through parts 1–4 inclusive.

It is important at this point to confirm that pupils are happy with
the idea of Truth tables, and the use of A, B, C etc. to denote
inputs and Z to denote an output.

Apparatus Required (15 pupils, 5 groups of 3)

five AND units
five NAND units
five OR units
five NOR units
five LED indicator units
five 5 volt stabilised voltage supply modules.
stackable 4 mm plug leads — assorted sizes.

Logic Assignment 4

In this assignment the pupils build logic gates from discrete
electrical and electronic components. They should realise that there
are many ways of building logic gates and that by using discrete
components they can switch motors on and off directly, whereas
with the integrated circuit logic units they would need to build a
buffer circuit to drive motors, 12 V lamps etc.

The circuits illustrated in the assignments are not the only ones
suitable for this work, but they have been chosen because they can
be built from the existing control technology equipment.

When the pupils have completed this part of the assignment they should be able to identify the type of gate they have made from comparison of Truth tables.

The pupils are then introduced to the block diagram notation of logic gates. The ones given are not universally accepted and so alternative symbols can be discussed.

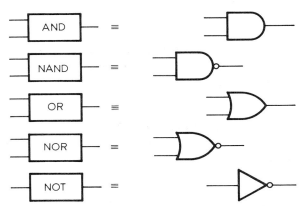

When other devices are introduced their symbols will be given. It should be demonstrated at this point that gates need not be limited to two inputs — three, four or even more are possible.

Apparatus Required (15 pupils, 5 groups of 3)

ten diode units,
ten transistor units,
ten 10 kilohm resistor units,
ten 1 kilohm resistor units (or fuse holder units fitted with resistors),
five LED indicator units,
five power supplies — 12 V, 2 A — with safety fuse or trip,
stackable 4 mm leads-assorted lengths.

Logic Assignment 5

This assignment introduces pupils to the idea that if they had for example, only NAND gates they could make up the other gates from these.

When they have completed parts 1 and 2, it would be helpful to show them other ways of combining, say, NAND gates to make AND, OR, and NOR gates.

Part 3 of this assignment brings the pupils back to the exclusive OR gate. This would be a suitable stage at which to discuss logic equations with the pupils. It is important that they understand the reasons for the use of brackets. E.g.

$$(A.\overline{B}) + (\overline{A}.B) = Z$$

meaning that (A AND NOT B) OR (B AND NOT A) are the conditions required for a '1' output. If the brackets were not included we would have

$$A.\overline{B} + \overline{A}.B = Z.$$

This could mean A AND (NOT B OR NOT A) AND B gives the required 1 output — which is nonsense. An analogy could be in the form of an algebraic expression. E.g.

$$(3x + 6) (2x - 3) = 10$$

This expression does not give the same answer as

$$3x + 6 (2x - 3) = 10$$

or

$$3x + 6.2x - 3 = 10$$

The solution included in the pupils books is only one way in which an exclusive or gate could be formed. Two other solutions are shown below.

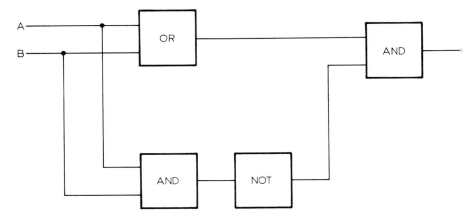

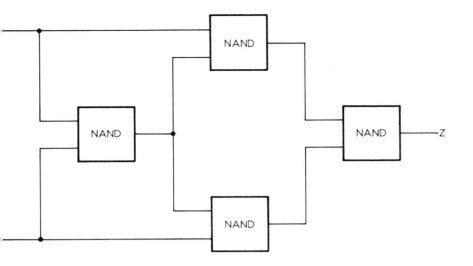

The final part of this assignment is a logical problem for the pupils to solve. It is suggested that 10 to 15 minutes is allowed in lesson time and that it should be completed for homework. It is suggested that pupils should build their electrical solutions.

Apparatus Required (15 pupils, 5 groups of 3)

Ten NOT units
Five NAND units
Fifteen NOR units
Ten AND units
Five OR units
Five LED Indicator units
Five 5 volt stabilised voltage supply modules
Five 5 power supplies — 12 V, 2 A — with safety fuse or trip
Stackable 4 mm plug leads — assorted sizes

A Possible Group Project _____

This project based on a turntable unit can be tried either after the
Transistor assignments or immediately following a short course on
logic and logic circuitry. Although some teachers will no doubt
have made use of a turntable as part of other projects from time to
time (involving different solutions by the various groups in a class),
this project is intended to involve all pupils in a class of about
fifteen. Each group of three or four pupils might be concerned
directly with some aspect of the problem. When all groups have
perfected their particular solution, all the various parts are brought
to produce a working device. *The theme of this project, and the
possible solution offered here, is intended only as a guide in order
that teachers and pupils will be encouraged to suggest and solve
problems along similar lines, usually working as a group.* Whilst
some teachers may wish to involve pupils in solving the problem
specified here, the solution offered is given primarily as a means of
showing how the project might be developed; it is certainly not the
only solution, and undoubtedly not the best, though teachers might
wish to make use of some of the ideas in one way or another with
pupils of high ability.

Definition of the Problem

The project is concerned with the building of a device which can
'recognise', and thereafter indicate the recognition of, the height
and diameter of cylindrically shaped objects placed upon the
turntable. Four cylinders are required: 75 mm × 25 mm dia.,
75 mm × 18 mm dia., 50 mm × 25 mm dia., and 50 mm ×
18 mm dia. Since these are to be placed upright on the moving turntable, mild steel is a suitable material. The weight of each is
sufficient to ensure that they remain in the upright position. Less
dense materials are more likely to 'topple' should there be any jerky
notion of the turntable.

The design of the circuitry for the control of the turntable and the
recognition of the articles placed upon it should be such that any of
the four objects can be identified individually, or any combination
indicated.

Since there are four different items, there are sixteen possible arrangements which can be accepted (or rejected) by the device. The arrangements which could be indicated as 'accepted' are:

1 none of them,

2 long and fat,

3 short and fat,

4 either short and fat or long and fat,

5 long and thin,

6 either long and thin or long and fat,

7 either long and thin or short and fat,

8 long and thin, short and fat, or long and fat (i.e. not short and thin).

9 short and thin,

0 either short and thin or long and fat,

1 either short and thin or short and fat,

2 short and thin, short and fat, or long and fat (i.e. not long and thin),

3 either short and thin or long and thin,

4 short and thin, long and thin, or long and fat (i.e. not short and fat),

5 short and thin, long and thin, or short and fat (i.e. not long and fat),

6 all of them.

For simplicity, 'long' refers to a 75 mm piece, 'short' to a 50 mm piece; 'thin' to a 18 mm dia. piece, and 'fat' to a 25 mm piece. In the solution offered, acceptance of any piece is indicated by the illumination of a lamp. For instance, if the device is 'set' for condition 12, the lamp will illuminate if a short and thin, short and fat, or long and fat cylinder is detected, but it will remain extinguished if a long and thin cylinder is detected. The versatility required from the device should now be evident, in that one must be able to 'program' it in any one of sixteen different ways. Teachers familiar with the principles of logic will no doubt recognise that here we have a logic problem. This does not necessarily mean, however, that pupils (or teachers) should have some knowledge of logic before attempting such a project; indeed, having found a solution which works, the advantage of a logical study of the problem becomes self-evident. A short course in logic followed by a reappraisal of the problem in logic terms could prove stimulating to the more able pupil.

Work Distribution

The problem as a whole can be conveniently divided into four parts, a group of, say, three pupils being responsible for each part. However, some coordination between groups is essential, since eventually each of the four parts must fit together to form a whole. Thus one group of three pupils could be responsible for this coordination and would probably need to spend quite a lot of their time working with all groups in some sort of sequence. A typical distribution could be:

a) a means of detecting height — short or long;

b) a means of detecting width — fat or thin;

c) a means of starting the turntable manually when a specimen has been placed upon the turntable, and a means of stopping it automatically after exactly one turntable revolution;

d) determining the number of possibilities and devising an 'accept' or 'reject' circuit when a cylinder is sensed;

e) coordination between the four groups.

Indication of 'accept' — say, lamp illuminated — and 'reject' — say, lamp remains extinguished — is probably best arranged to occur at the end of the cycle when the turntable stops.

A Possible Solution

This solution is based entirely on the basic units designed specially for the course. Height and width are detected using photocells and associated light sources, the photocells being mounted at the periphery of the turntable unit and the light sources at the centre. In order to do this, a gantry spanning the unit is required; the light sources are then attached to a vertical dropped from the centre of the gantry. The 'accept' or 'reject' indicator lamp is also placed on the gantry. Connecting leads for the indicator and light sources should be attached to the gantry to prevent any obstruction of the cylinders as they rotate on the turntable.

Fig.1

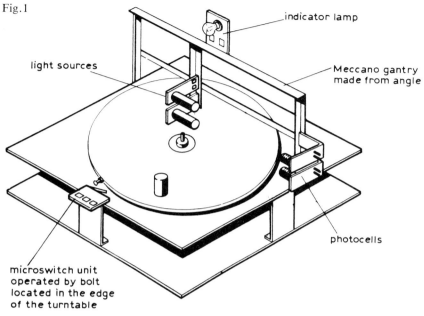

indicator lamp

light sources

Meccano gantry made from angle

photocells

microswitch unit
operated by bolt
located in the edge
of the turntable

A hole is drilled into the edge of the turntable proper; the hole is tapped M3; and a screw is part-way inserted and locked in position with a nut. This screw operates a microswitch attached to one of the existing Meccano brackets which space the top and bottom square Perspex turntable plates.

ig. 2

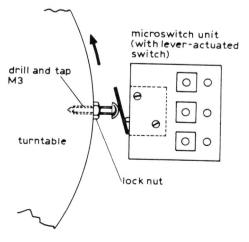

microswitch unit
(with lever-actuated
switch)

drill and tap
M3

turntable

lock nut

Control Unit and Turntable-Motor Control

Control of the turntable motor is achieved through a relay bistable unit. This unit is triggered either by the microswitch on the turntable or by the manual push button circuit (control unit). The connections C, in conjunction with a power supply, two sets of switch contacts (A and B) and a lamp indicator unit, enable any of the sixteen different arrangements to be selected. The wiring between points S is dealt with later. In addition to triggering the bistable unit, the control unit receives and sends information to the height-detection and width-detection units. This information is carried through connections W, X, and Y. When the manual push button on the control unit is pressed momentarily, the short-circuit across the terminals 1 and 2 in the bistable unit causes it to change over; the turntable motor starts to run; and a short-circuit appears across the connections C, when it eventually stops. Because of the latter occurrence, the indicator on the gantry illuminates under the correct conditions only at the end of a cycle. Note that the manual push button must be pressed for a sufficient time to allow the bolt-head to break contact with the microswitch situated at the periphery of the turntable.

NOTE: In this and all other circuits relating to the project, separate power supplies are shown in each unit, to preserve clarity. In practice, only one is necessary.

Height Detection

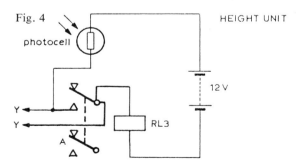

Fig. 4 HEIGHT UNIT

photocell

12 V

Y

Y

RL3

A

The height detector makes use of the photocells uppermost on the gantry, in conjuction with a two-pole, two-way relay unit. Since the photocell is normally illuminated — only a long specimen prevents light striking this cell — the relay will be normally energised.

Fig. 3

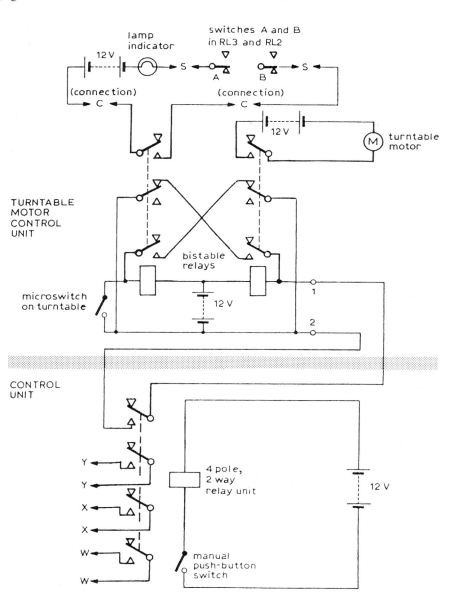

Because of the circuit arrangement, however, the relay can energise only if the connections Y are short-circuited, this occurring briefly when the manual push button on the control unit is operated. If a long specimen prevents light striking the photocell, the relay de-energises and remains de-energised until the start of the next test, when the manual push button on the control unit is again depressed. Thus the device 'remembers' it has just sensed a long specimen. Alternatively, if the photocell receives light throughout the whole cycle, it 'remembers' that either the specimen was short or that there was no specimen present. The set of switch contacts marked A, mentioned previously under 'Control unit and turntable motor control', will be dealt with later.

Width Detection

Width detection is described using two different techniques; the first uses a transistor circuit and the second a simpler arrangement similar to the height-detection unit. Although it is more involved, the transistor design is more reliable than the simpler relay circuit, and it is more easily adjusted. The bottom photocell is used for width detection. Both long and short specimens are arranged to block the light reaching the photocell, but obviously a fat (large diameter) specimen will block the light to the cell for a longer time than a thin one. Both detection methods are based upon this difference in time.

Width Unit Transistor Arrangement

RL_1 is always energised when light is incident upon the photocell, and under this condition the base circuit (50 kilohm variable) of the transistor is disconnected from the power supply and is inoperative. When any specimen, thin or fat, breaks the beam to the photocell, RL_1 immediately de-energises and supply is fed to the transistor base. The transistor does not immediately conduct, however, since a large value capacitor (2500 μF approx.) is connected between the base and emitter. This capacitor is initially uncharged and behaves like a short-circuit; thus instead of current flowing to the transistor base, it charges the capacitor. As the capacitor charges, its effective resistance gradually increases and an increasing base current begins to flow. Initially RL2 is energised, it is connected to the supply through the 50 ohm resistor, and the transistor is not conducting.

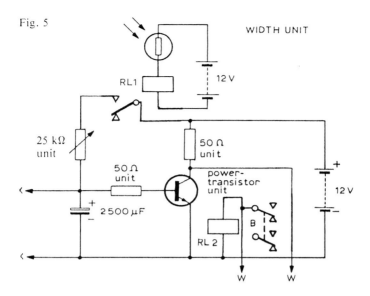

Fig. 5

WIDTH UNIT

RL1

12 V

25 kΩ
unit

50 Ω
unit

50 Ω
unit

power-
transistor
unit

12 V

2500 μF

RL 2

B

W   W

As the transistor begins to conduct, however, its falling resistance causes the voltage, across the transistor to fall. This voltage is also across the relay coil (since they are both in parallel) and, if the photocell remains non-illuminated for a long period (fat specimen), the voltage across the relay falls to too low a value to keep it energised. It therefore changes over to the de-energised state until the contacts W are short-circuited by the manual push button; thus the device 'remembers' that a fat specimen has been detected. A thin specimen breaks the light to the photocell for a shorter duration, and the current in the transistor does not reach a sufficiently high value to case the relay RL2 to de-energise. Obviously critical adjustment is necessary, and this is achieved through the 50 kilohm variable resistance which changes the rate at which the capacitor can charge. If the relay RL2 de-energises even when a thin specimen passes, then the resistance of the variable resistor must be increased to cause a longer delay. The contacts B have been mentioned previously under 'Control unit and turntable-control motor', and will be dealt with later.

Width Unit, Simple Arrangement:

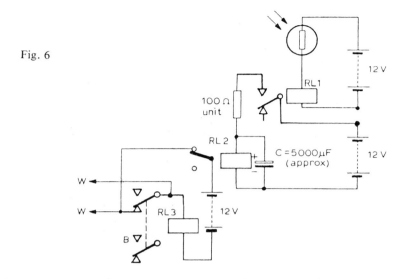

Fig. 6

This relay circuit is only one of a number of possibilities. RL1 is energised whenever light is incident upon the photocell. When the beam is broken, supply is fed to RL2 through the 100 ohm variable resistor. However, RL2 does not immediately energise owing to the uncharged capacitor C (5000 μF approx.). This capacitor behaves as a very low resistance in its uncharged state, causing the bulk of the current to by-pass the relay coil. As the capacitor charges, a greater proportion of the current flows in the relay coil until there is sufficient to energise it. The relay time is determined by the setting of the variable resistor — the greater the resistance, the slower the charging rate. It should be noted, however, that there is likely to be a setting beyond which the relay will never energise, too large a resistance limiting the maximum current to a value which is too small to energise the relay — even when the capacitor has fully charged. Adjust the variable resistor until RL2 energises for a fat specimen but not for a thin.

If RL2 energises, the normally energised RL3 de-energises and remains de-energised irrespective of any subsequent state of RL2. Thus the circuit 'remembers' if a fat specimen has been detected. Contacts B are dealt with later.

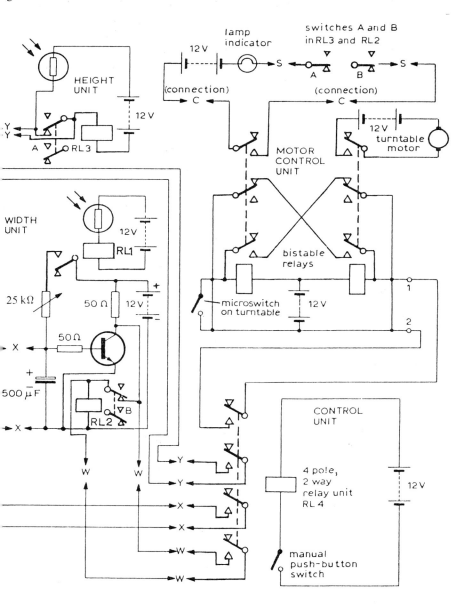

Selection of Function (programming)

As mentioned previously, the selection of any of the sixteen possible arrangements is achieved by suitably wiring the two connections S, a power supply, the indicator lamp on the gantry, the switch contacts A in the height-detection unit and the switch connections B in the width-detection unit. The switching arrangement necessary for any particular function is inserted between points S in the common circuit of fig. 8. The wiring of switches A and B for each function is shown in table 1.

Fig. 8

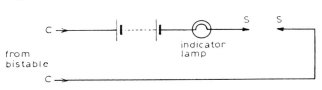

TABLE 1

Selection	Arrangement of A and B between S-S
1 None	
2 Long and fat	
3 Short and fat	
4 Either short and fat or long and fat	
5 Long and thin	
6 Either long and thin or long and fat	
7 Either long and thin or short and fat	
8 Long and thin Short and fat Long and fat	
9 Short and thin	
10 Either short and thin or long and fat	

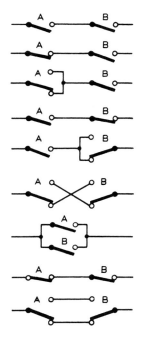

1 Either short and thin
  or short and fat

2 Short and thin
  Short and fat
  Long and fat

3 Either short and thin
  or long and thin

4 Short and thin
  Long and thin
  Long and fat

5 Short and thin
  Long and thin
  Short and fat

6 All

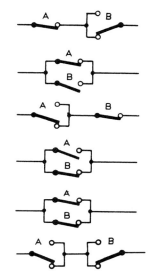

Note that the position shown for switches A and B are those prior to sensing.

Example:

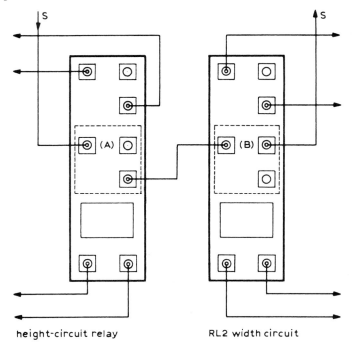

height-circuit relay   RL2 width circuit   221

Testing

1 Test each unit separately and to ensure that it operates correctly, simulate the working conditions. Temporarily connect a lamp and power supply to each of the contacts A and B in order to determine the states of the relays during testing.

Height Unit

Switch on the supply. The relay should not energise. Short together the Y connections and the relay should energise; if not, check that light is striking the photocell. If the relay energises, break the beam to the photocell and the relay should de-energise and remain de-energised. Again, short together the connections Y and the relay should energise. This completes the testing of the height unit.

Width Unit — Transistor Type

Turn the 25 kilohm variable resistor to minimum resistance. Switch on the supply. Both relays should immediately energise. If the photocell relay (RL1 — complete circuit diagram) does not energise, check that light is striking the photocell. If RL1 energises but RL2 does not, check the transistor-circuit wiring. If both relays energise, break the beam of light to the photocell and both relays should de-energise. Allow light to strike the cell again; RL1 should immediately energise and RL2 should remain de-energised. Short-circuit the connections W, and RL2 should again energise.

Rotate the 25 kilohm variable resistor about ¼ turn. Break the light beam to the photocell momentarily. RL1 should de-energise but RL2 should remain energised. Now break the beam for a much longer time and both relays should de-energies. Short together both connections W and connections X, and RL2 should again energise. This completes the testing of the width unit.

Control Unit

This simple unit should give no problems. Switch on the supply and push the manual push button switch. The relay should energise when the switch is depressed and de-energise when the switch is released.

Motor-Control Unit

Connect a lamp and a power supply in series with the connections C. Alternately shorting out the terminals marked 1 and 2 and operating a microswitch fitted across the other pair of input terminals should cause the bistable to change state.

Interconnect each unit. Wire in the turntable microswitch and motor. Press the manual push button and the turntable should rotate 1 revolution, stopping again when the turntable microswitch is operated by the rotating turntable. Wire up connections C in series with the power supply and the indicator lamp on the gantry. This should leave two flying leads S. Connect one of these leads to the pole of relay switch contacts A on the height unit and the other to the normally open (rear) contact. Place a tall specimen on the turntable. Press the manual push button, allowing the turntable to rotate one revolution. When it stops, the indicator should illuminate, showing that it has recognised a tall object. Repeat for a short specimen and the lamp should not illuminate.

Remove the flying leads S from relay contacts A on the height unit and wire them similarly to the relay contacts B on the width unit. Place the short and thin specimen on the turntable and press the manual push button switch. When the revolution is complete, the indicator lamp should remain extinguished. Repeat for the short and fat specimen and the lamp should illuminate. Find a setting of the 25 kilohm variable resistor which causes the lamp to illuminate with a short and fat specimen but not with a short and thin specimen. The testing is now complete. By suitable wiring of contacts A and B, as shown in table 1, any selection of specimens should be possible.

An Explanation Using the Rules of Logic

This section is intended for teachers who are familiar with logic and logic circuitry. Some of these teachers may wish to discuss the project after completion, along the following lines with the more able pupils.

There are sixteen alternative selections when using four different 'characteristics' — four different 'characteristics' being short, long, thin, and fat. Using binary notation, the sixteen are represented as shown in table 2.

TABLE 2

Note: '1' represents a selection; 'O' represents no selection.

	Short thin	Short fat	Long thin	Long fat
Line 1	0	0	0	0
Line 2	0	0	0	1
Line 3	0	1	0	0
Line 4	0	1	0	1
Line 5	0	0	1	0
Line 6	0	0	1	1
Line 7	0	1	1	0
Line 8	0	1	1	1
Line 9	1	0	0	0
Line 10	1	0	0	1
Line 11	1	1	0	0
Line 12	1	1	0	1
Line 13	1	0	1	0
Line 14	1	0	1	1
Line 15	1	1	1	0
Line 16	1	1	1	1

The first line means that we must provide for the selection of none of the specimens; the second long and fat; the fourth either short and fat or long and fat; the eighth long and thin, short and fat, or long and fat; and so on.

One can now write out truth tables for each 'line' in which the switches A and B represent the 'inputs' of 1 or 0 (operated or not operated) and each line represents the output column — note that our horizontal lines now become vertical columns. The output columns will be labelled Z (table 3). Alongside each truth table is the logic equation in Boolean algebraic form, the selection which causes the indicator lamp to illuminate, and the wiring of switches A and B to give this selection. To simplify the description, short and thin has been abbreviated to ST, long and thin to LT, short and fat to SF, and long fat to LF. The switching arrangement necessary for any particular selection is inserted between points S in fig. 9.

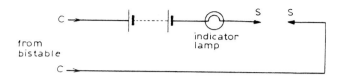

TABLE 3

Line in table 2	Truth table			Logic equation	Selection	Wiring between S — S
1) 0000	A	B	Z		None	Omitted
ST	0	0	0			
SF	0	0	0			
LT	0	0	0			
LF	0	0	0			
2) 0001	A	B	Z	A. B = Z	Long and fat	
ST	0	0	0	(AND)		
SF	0	1	0			
LT	1	0	0			
LF	1	1	1			
3) 0100	A	B	Z	B. $\overline{A}$ = Z	Short and fat	
ST	0	0	0			
SF	0	1	1			
LT	1	0	0			
LF	1	1	0			
4) 0101	A	B	Z	(A. B) +	Either short	
ST	0	0	0	B = Z	and fat or	
SF	0	1	1		long and fat	
LT	1	0	0			
LF	1	1	1			
5) 0010	A	B	Z	A. $\overline{B}$ = Z	Long and thin	
ST	0	0	0			
SF	0	1	0			
LT	1	0	1			
LF	1	1	0			

225

6)	0011	A	B	Z	(A. B)	Either long
	ST	0	0	0	v A = Z	and thin or
	SF	0	1	0		long and fat
	LT	1	0	1		
	LF	1	1	1		

7)	0110	A	B	Z	A + B.	Either long
	ST	0	0	0	A. B = Z	and thin or
	SF	0	1	1	(Exclusive	short and fat
	LT	1	0	1	OR)	
	LF	1	1	0		

8)	0111	A	B	Z	A + B	Long and
	ST	0	0	0	= Z (OR)	thin, short
	SF	0	1	1		and fat,
	LT	1	0	1		long and fat
	LF	1	1	1		

9)	1000	A	B	Z	A + B	Short and thin
	ST	0	0	1	= Z	
	SF	0	1	0	(NOR)	
	LT	1	0	0		
	LF	1	1	0		

10)	1001	A	B	Z	A. B +	Either short
	ST	0	0	1	A + B	and thin or
	SF	0	1	0	= Z	long and fat
	LT	1	0	0	(Logical	
	LF	1	1	1	Equiva-	
					lence)	

11)	1100	A	B	Z	(A. B) +	Either short
	ST	0	0	1	A = Z	and thin or
	SF	0	1	1		short and fat
	LT	1	0	0		
	LF	1	1	0		

12)	1101	A	B	Z	B + A +	Short and
	ST	0	0	1	B = Z	thin, short
	SF	0	1	1		and fat, long
	LT	1	0	0		and fat
	LF	1	1	1		

13) 1010	A	B	Z	(A . B) +	Either short
ST	0	0	1	B = Z	and thin or
SF	0	1	0		long and thin
LT	1	0	1		
LF	1	1	0		

14) 1011	A	B	Z	B . $\overline{A}$ = Z	Short and
ST	0	0	1		thin, long
SF	0	1	0		and thin,
LT	1	0	1		long and fat
LF	1	1	1		

15) 1110	A	B	Z	A . B = Z	Short and thin
ST	0	0	1	(NAND)	Long and thin
SF	0	1	1		Short and fat
LT	1	0	1		
LF	1	1	0		

16) 1111	A	B	Z	A + B +	All
ST	0	0	1	A +	
SF	0	1	1	B = Z	
LT	1	0	1		
LF	1	1	1		

Appendix 1

Photoelastic Stress Analysis

The phenomenon of photoelasticity makes it possible to measure stress and strain by detecting changes in the indexes of refraction of light passing through the photoelastic material. The principle is based on the fact that polarised light passing through a transparent, birefringent plastics material under strain will split into two polarised beams which travel in the planes of the principal strains. These beams will have different velocities and therefore different wavelengths, and the result is the production of a series of coloured areas or lines. These are known as 'isochromatics', in which areas of equal colour represent areas of equal stress. In this range of colours, there is a distinct zone where the red is immediately followed by a blue-green. This zone, which can easily be located, is known as the 'tint of passage' or fringe boundary.

The polarising axes of the polariser and analyser are at right angles

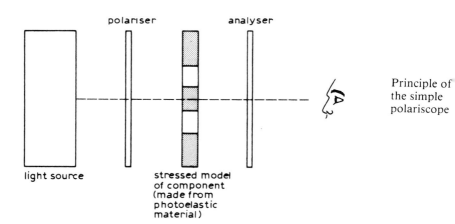

Principle of the simple polariscope

If the material is stressed sufficiently, there will be more than one complete sequence of colours, and several fringes will appear, the number of fringes being a linear measure of the magnitude of the stress.

Providing the polarisers used are large enough, the whole of the model of the component can be examined. This ability to see the

complete stress field is one of the important advantages of the
photoelastic technique.

Principle of the polariscope

The design of a simple polariscope required for the photoelastic
demonstrations can be based on the items shown here.

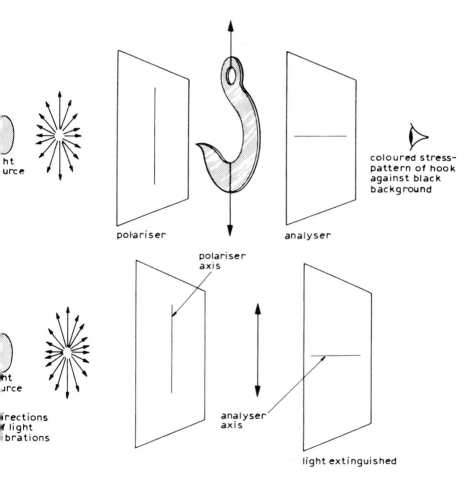

To demonstrate the forces present in simple structures, it is often
necessary to compare one design solution with another. It is very
difficult to remember (or record by sketching) the pattern produced
in one experiment and compare this with another produced just a

few seconds later. It is suggested, therefore, that specimens should be mounted side by side for easy comparison. It is possible to obtain very large polarisers (200 mm × 200 mm or larger), but smaller ones are obviously cheaper and easier to obtain, and, if these are used, a double side-by-side polariscope is recommended.

Since demonstrations will be given to fairly large groups, it is important to keep the polariser and analyser as close together as possible, to provide a wide field of view.

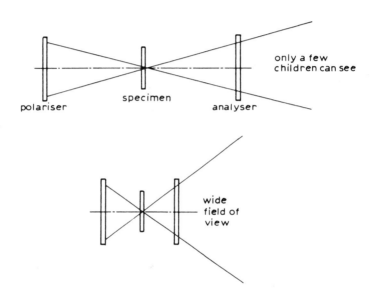

However, if the polariser and analyser are close together, it is very difficult to position the specimen. It is therefore suggested that the analyser and the straining frame should be easily removable from the polariscope.

Construction of a straining frame

The straining frame must provide a wide range of fixing positions, and for this reason Meccano is very suitable.

Each specimen will demand a different form of support, so it is impossible to be specific on the design of this part. The Meccano framework shown should, with the addition of a few extra members, provide a means of holding most specimens.

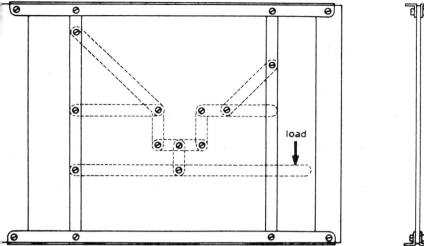

Shaping the specimen

Great care must be taken to ensure that the sheet is not internally stressed during the shaping process. Do not hold the sheet in a vice at any time. Cut with a bandsaw (ensuring that the sheet does not become hot) or with a hacksaw, holding the sheet flat on a bench by hand. Hole are best drilled by end-milling cutters. The specimen must be made with great accuracy if valid comparisons are to be made.

Possible Alternatives

Schools possessing a Polaroid camera or having other resources for photography will be able to make photographic records of stress patterns. In this case, only one model of the specimen need be made, and only one polariser/analyser pair will be required.

Literature, obtainable from CIBA (UK) Ltd, Duxford, Cambs, describes the production of models in Araldite resin which is highly sensitive to small changes in stress. Moreover, the stress pattern can be 'preserved' by baking the model together with the loading frame. On cooling, the model can be viewed without the need to reload.

Suppliers of equipment and further information

Polarisers (UK) Ltd. Lincoln Road, Cressex Estate, High Wycombe, Bucks.

Stress Engineering Service Ltd., Charlton Lane, Midsommer-Norton, Bath. BA3 4BE

Sharples Photomechanics Ltd., Europa Works, Wesley Street, Bamber Bridge, Preston PR5 4PB

Welwyn Strain Measurement Ltd., Armstrong Road, Basingstoke. RG24 1QA

For further guidance on photoelastic techniques see 'Photoelasticity for Schools and Colleges' published by NCST, Trent Polytechnic, Nottingham.

Appendix 2

Stain-Gauges

The use of strain-gauges on model structures

The resistance strain-gauge is a useful device for determining the magnitude and the kind of force (compressive or tensile) present in structural members. A high-gain amplifier is required to amplify the effect of the small changes which take place in the resistance of a strain-gauge attached to a member in tension, compression, or bending.

It is not practicable to use strain-gauges on Meccano strip for two reasons:

a) Meccano strip is easily bent, and may be appreciably so even when new;

b) the presence of holes in the material results in a complex force distribution when the material is loaded.

For these reasons, if it is desired to obtain the relative sizes of the forces in members or, more simply, to determine whether or not a member is in tension or compression, the relevant Meccano member is removed and replaced by a length of strip or angle Perspex (3 mm in thickness) to which a suitable strain-gauge has been

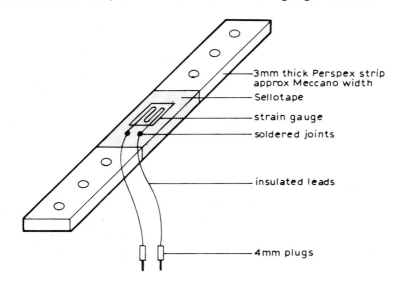

3mm thick Perspex strip approx Meccano width

Sellotape

strain gauge

soldered joints

insulated leads

4mm plugs

attached. It is not necessary to have Perspex replicas of every size of Meccano angle and strip. Provided that no holes are present within about 25 mm either side of the strain-gauge, a long strip can be made and used to replace any length of Meccano. In most cases the resulting projection of Perspex from the model will not invalidate the test.

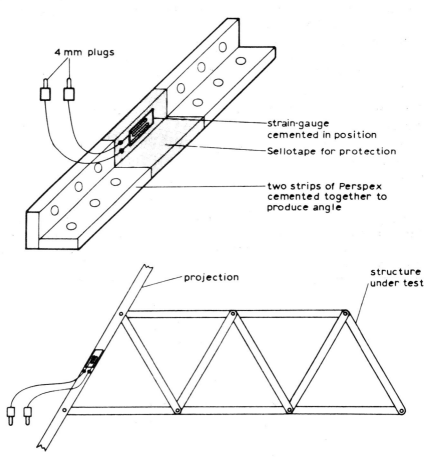

It is recommended that, if a test is to be made to determine whether a member is in tension or compression, angle material made from Perspex is used, since this will not bend under compression. When tests are made to find the relative size of forces in members, it must be remembered that only members of the same section can be compared.

Fixing strain-gauges

Clean the Perspex with a degreasing agent which will not attack the material e.g. methylated spirit. Alternatively, use very fine glass paper or emery cloth over an area slightly larger than the gauge. Apply a thin layer of quick-drying adhesive (type CH — supplied by Electro Mechanisms Ltd) to the back of the gauge, place the gauge in position on the Perspex, cover with a small piece of poly-ethylene sheet, and press in place betwen the fingers for about one minute. Remove the polyethylene sheet and leave to set fully for about thirty minutes.

Cut the gauge leads, leaving about 6 mm of free lead, and carefully solder on to each an extension lead of insulated wire about 500 mm long. Secure the extension leads to the Perspex and protect the gauge by wrapping Sellotape tightly round the member. Fit 4 mm plugs on the extension lead.

Resistance Changes in Strain-Gauges

If the gauge is fitted as shown, the following changes in resistance occur when the member is in tension, compression or bending.

a) Member in tension — resistance increases.

b) Member in compression — resistance decreases.

c) Member in bending — top in tension, resistance increases.

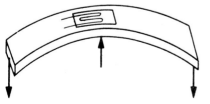

d) Member in bending — top in compression, resistance decreases.

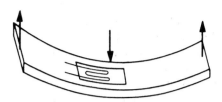

From these diagrams it is apparent why a single Perspex strip should not be used to determine whether a member is in tension or compression. If the member bends when in compression it could do so in either direction, thus it is possible to obtain an increase in resistance (diagram d) suggesting that the member is in tension.

Strain-Gauge Amplifier

A high-gain d.c. amplifier free from appreciable 'drift' is a complex and expensive item, but in recent years schools have constructed low-cost instruments based on readily available (integrated circuit) operational amplifiers.

For guidance on the construction of a strain-gauge bridge using an operational amplifier, refer to Equipment Guide page 92, published by NCST, Trent Polytechnic, Nottingham.

Appendix 3 _____

Capacitors

The pupils will have used a capacitor in conjunction with a relay in the Electrical-switching assignments in order to produce a 'delay device'. The simple theory of capacitors is probably best covered by simple experiments, working in groups, and by teacher demonstrations. The following procedure should be adopted.

1   Discuss the effects when using the delay circuit in the vehicle. The idea of a capacitor acting as a short-term battery or reservoir may well be put forward. Emphasise this 'delay' function, since the capacitor is included as one of the three main devices which can be used in conjunction with logic arrangements which may come later.

These arrangements are:

the clock (oscillator Horastable),
the delay (monostable) , and
the flip-flop (bistable).

2   Each group is asked to connect a capacitor (about 5000 μF) across the terminals of a 12 V battery. The supply is removed and the capacitor terminals are then shorted out. Discuss the implications of a spark appearing. The storing of an electrical charge should be emphasised, and it may be wise at this stage to again discuss the difference between 'charge' and 'current'.

3   'What determines the amount of charge stored?'
The pupils should be encouraged to try different voltages and different values of capacitor (not capacitors in parallel to increase the value at this stage). Since the size of spark gives and indication of the quantity of charge stored, they should come to the conclusion that:

a) the larger the voltage used, the greater the charge stored; and

b) the larger the value of the capacitor, the greater the charge stored.

With able pupils we could suggest:

$$Q \propto C$$
$$Q \propto V$$
$$\therefore Q \propto CV$$

The units of capacitance — the farad (F), microfarad (μF), and picofarad (pF) — should be explained. The main point to be made is that the larger the number of microfarads, the larger the storage capacity of the capacitor.

4 Pupils should then be encouraged to use capacitors in parallel, a 12 V d.c. supply. It will be seen that the capacitance increases if two or more capacitors are connected in parallel. Hence two 5000 μF capacitors in parallel produce a total capacitance of 10 000 μF, three 5000 μF capacitors in parallel give 15 000 μF, and so on.

5 The effect of capacitors in series could be investigated, but the application of this arrangement in project work is likely to be very limited during the course.

6 Teachers may wish to discharge a capacitor through an ammeter to show current flow, to reinforce the spark demonstration. Capacitors of about 5000 μF charged up using a 12 V d.c. supply produce adequate deflections on a 0–1 A ammeter. It should be appreciated, however, that the inertia of the moving parts of the ammeter may result in very low values of current being indicated.

7 The delay function of a capacitor is an important item in the course and should, therefore, be treated quite fully. If a capacitor is charged from a battery, then removed from the supply and discharged by placing a piece of wire across the capacitor leads, it can be seen that the charge and discharge are virtually instantaneous. However, if the charge or discharge operation is carried out with a resistance in series with the capacitor (e.g. a resistor or a relay coil) the charge and discharge times are much slower. This effect should be shown using resistors, since the pupils will have noticed the effect using a relay. We must show that the resistance of the relay coil is related to the delay time produce. The following experiments are suggested:

a) Connect a 12 V, 2.2 W bulb across a 12 V supply. Remove one lead from the supply and note how rapidly the bulb is extinguished. Charge a 5000 μF capacitor from a 12 V supply (d.c.). Discharge it through the same bulb and note the slight delay before the bulb finally goes out.

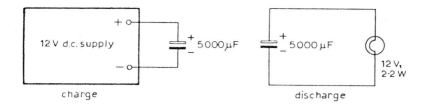

charge   discharge

b) Charge a 5000 μF capacity from a 12 V d.c. supply.
Discharge the capacitor through a robust moving-coil
ammeter (0–1 A). Note the rapid discharge.

Note: A charge instantaneous current will flow initially, and
this could damage a fragile meter.

Place a resistor of about a 2000 ohms in series with a
0–10 mA meter (the 0–10 mA range on an Avometer model
7 is suitable). Note the slow charge and discharge rates.
Draw graphs of charging current plotted against time and
discharging current against time, with the aid of a
stopwatch.

From graphs such as these, we can calculate the delay time
which will be obtained with a 5000 μF capacitor in parallel
with a relay of coil resistance 2000 ohm, when the minimum
holding current of the relay is known.

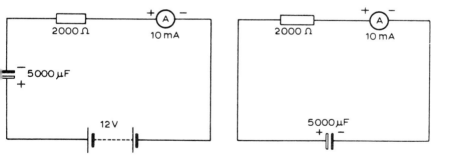

Discuss the results of these experiments and indicate how the
delay time for a capacitor discharging through a relay is
mainly dependent upon the resistance of the relay coil. To
emphasise this fact, it is recommended that a number of
relays, with different coil resistances, be tested and the delay
times found.

8   Show a number of different types of capacitor, relating physical size and type to capacitance. The importance of correct polarity when using electrolytic types could be discussed and a comparison made with other types.

Appendix 4

Course Equipment

Background

Throughout the development of the CONTROL TECHNOLOGY Course it was recognised that much of the work to be carried out by the pupils, and the teaching methods to be adopted, should be based, primarily, on the use of a range of purpose-built equipment. In devising the apparatus which would effectively complement the pupils' texts, careful consideration was given to factors such as availability, costs, appearance, practicality and durability.

A policy of continuous development and modification of the basic equipment was adopted during the period of trials, and later, thus reflecting the day-to-day experience in the schools. Not only was the equipment to be of central importance for the boy or girl working sequentially through the programme of assignments and investigations, but it was to be used extensively in project work at various stages in the course.

With all of these aspects in mind, and conscious of the various pressures on the schools, negotiations have taken place with a number of manufacturers which have resulted in substantial savings to those schools placing orders for complete kits. In addition, agreement has been reached regarding simplified ordering procedures, thereby avoiding the frustrating use of extensive and complicated parts lists.

Up-to-date information on the provision of the complete equipment for the course is available from:

The National Centre for School Technology,
Trent Polytechnic,
Burton Street,
Nottingham.